Advances in Exploration, Development and Utilization of Coal and Coal-Related Resources

Advances in Exploration, Development and Utilization of Coal and Coal-Related Resources

Editors

Jing Li
Yidong Cai
Lei Zhao

MDPI • Basel • Beijing • Wuhan • Barcelona • Belgrade • Manchester • Tokyo • Cluj • Tianjin

Editors
Jing Li
China University of
Geosciences (Wuhan)
China

Yidong Cai
China University of
Geosciences (Beijing)
China

Lei Zhao
China University of Mining
and Technology (Beijing)
China

Editorial Office
MDPI
St. Alban-Anlage 66
4052 Basel, Switzerland

This is a reprint of articles from the Special Issue published online in the open access journal *Energies* (ISSN 1996-1073) (available at: https://www.mdpi.com/si/energies/AEDU_CCR).

For citation purposes, cite each article independently as indicated on the article page online and as indicated below:

LastName, A.A.; LastName, B.B.; LastName, C.C. Article Title. *Journal Name* **Year**, *Volume Number*, Page Range.

ISBN 978-3-0365-6191-2 (Hbk)
ISBN 978-3-0365-6192-9 (PDF)

Cover image courtesy of Dr. Jing Li

Contents

About the Editors

Jing Li

Jing Li is an Associate Professor at the China University of Geosciences, Wuhan. Her research fields include coal geology; critical elements in coal and coal combustion products (CCPs); coal geochemistry; coal mineralogy; the environmental geochemistry of CCPs; and the sustainable and high-value-added utilization of CCPs.

Yidong Cai

Yidong Cai is a Professor at the China University of Geosciences, Beijing. He is an expert on unconventional resources evaluation; pore/fracture structure; fluid transport; and micromechanical properties.

Lei Zhao

Lei Zhao is a Professor at the China University of Mining and Technology (Beijing). She is famous for her research on mineral matter and trace elements in coal; strategically critical metals in coal-bearing sequences and coal combustion byproducts; and the environmental impact of coal combustion by-products.

Preface to "Advances in Exploration, Development and Utilization of Coal and Coal-Related Resources"

"Advances in Exploration, Development and Utilization of Coal and Coal-Related Resources" have been the research frontier and hot topic in coal geology, and are of great significance to achieving carbon neutrality. This book encompasses a series of representative advanced research on geological and geochemical characterization, as well as the exploration, development, and utilization of coal and coal-related resources, especially critical metals together with unconventional gases in coal and coal-bearing sequences.

Jing Li, Yidong Cai, and Lei Zhao
Editors

Editorial

Advances in Exploration, Development and Utilization of Coal and Coal-Related Resources: An Overview

Jing Li [1,*], Yidong Cai [2] and Lei Zhao [3]

[1] Key Laboratory of Tectonics and Petroleum Resources, China University of Geosciences, Wuhan 430074, China
[2] School of Energy Resources, China University of Geosciences, Beijing 100083, China
[3] School of Earth Sciences and Surveying and Mapping Engineering, China University of Mining and Technology (Beijing), Beijing 100083, China
* Correspondence: jingli@cug.edu.cn

Citation: Li, J.; Cai, Y.; Zhao, L. Advances in Exploration, Development and Utilization of Coal and Coal-Related Resources: An Overview. *Energies* **2022**, *15*, 9304. https://doi.org/10.3390/en15249304

Received: 1 November 2022
Accepted: 30 November 2022
Published: 8 December 2022

Publisher's Note: MDPI stays neutral with regard to jurisdictional claims in published maps and institutional affiliations.

The worldwide development of clean and low-carbon energy is undoubtedly imperative in the coming decades. Coal is definitely an indispensable source of energy necessary for techno-economic progress, currently accounting for 41% of global electricity needs. The combustion of a high volume of coal has led to the release of CO_2, NO_x, SO_2 and toxic elements, as well as to the accumulation of huge amounts of coal combustion products (CCPs), which may pose a severe threat to ecosystems [1]. As a consequence, deep exploration and development, as well as the clean and efficient utilization of coal and coal-related resources (e.g., toxic elements in coal, strategic metals in coal, coal measure gases, black shale gases, etc.), is of great significance to achieve carbon neutrality.

As is known, strategic metal resources are extremely important in economic development and defense security. Coal is a particular sedimentary organic rock, in which strategic metals (e.g., Ge, U, Ga, Li, and rare earth elements and Y) can be enriched and subsequently form large or super-large ore deposits under some specific geological conditions [2,3]. Given the continuous discovery and promising prospect of strategic metal ore deposits in coal-bearing strata, research on coal-hosted strategic metal ore deposits has been one of the most important subjects in coal geology and mineral deposit, and also been the frontier issue of international mineral geology. The recovery of critical metals from coal and CCPs will contribute not only to the remission of global demand for strategic metal resources but also to the clean and efficient utilization of coal. The novel utilization of CCPs, e.g., synthesis of zeolite, manufacture of foam glass and fire-resistant panels is not only eco-friendly (saving useful land, reducing emissions of pollutants and greenhouse gas), but also economically significant [4–8]. However, due to the complexity of the necessary questions and difficulties, the research on ore-forming theory, exploitation and exploration, and extraction technologies of strategic metals, as well as on development of new technology for the efficient utilization of CCPs, still faces big challenges and needs further investigation.

Meanwhile, research on the characterization and evaluation of coal measure gases, and other coal-related unconventional resources in coal and coal-bearing sequences, will also greatly promote the development of low-carbon energy. For instance, research on controlling factors of coal measure gas production and on advanced techniques to increase coal measure gas production, is to a large extent conducive to achieving carbon neutrality. In addition, natural gas generation is the result of organic matter degradation under the effects of biodegradation and thermal degradation. Black shales are rich in organic matter and have shown great shale gas potentiality in recent years. Note that stone coal, a special kind of coal, is a combustible, low-heat value, high-rank black shale mainly derived from early Paleozoic bacteria and algae after saprofication and coalification in a marine-influenced environment [9]. Apart from the great potentiality of shale gas, under some special geological conditions, stone coals are also enriched in strategic elements (V and Se) that are industrially and agriculturally utilized, as well as some toxic elements that have caused environmental pollution (e.g., SO_2 emission during their combustion) and endemic diseases such as selenoisis and fluorosis. Furthermore, research on stone coal can provide useful information for geological events and regional geological setting (e.g., hydrothermal activities).

Furthermore, the exploration and mining process of coal seams, especially those thin and medium thick coal seams with complex geological structure and close distance, is mostly accompanied by violent mine pressure and continuous large deformation of surrounding rock, which will restrict the improvement of the coal mine safety situation and production efficiency in coal mines. Therefore, it is urgent to further reveal the deformation characteristics of surrounding rock and to provide the scientific basis for safe exploration and mining process of coal seams.

The Special Issue encompasses a series of research papers, which provide several representative research aspects of advances in exploration, development and utilization of coal and coal-related resources. For instance, Li et al. and Wei et al. discussed the geological and geochemical characterization, respectively, of coals from typical coalfields in China, with special emphasis on the strategic metal enrichment in coal [10,11]. Zhang et al. investigated the distribution, occurrence, and integration of As in Permian coals in North China, and inverted influence of depositional environment on enrichment of As and other toxic elements in coal [12]. These studies will play an important role in the potential extraction of strategic metals from coal, as well as in environmental protection (environmental pollution control and the implementation of carbon neutralization).

With respect to the advanced research on the characterization and evaluation of coal measure gases to fulfill the efficient exploration, development and utilization of coal measure gases, Chen et al. revealed the controlling mechanism of fracturing fluid on CBM migration and found that both fracturing fluid volume and ratios of critical desorption pressure to reservoir pressure (Rc/r) have a significant impact on coal measure gas production [13]. This research can provide guidance for the optimal design of hydraulic fracturing parameters for CBM wells. With respect to the research on characterization and evaluation of other unconventional resources, Xia et al. conducted a comprehensive study of marine redox conditions, primary productivity, sedimentation rate, terrigenous input and hydrothermal activity of early Cambrian Niutitang black shales (usually called stone-coal) in the Upper Yangtze Region, and revealed the organic matter enrichment of the black shales deposited in different settings [14]. This study provides important information for predicting favorable areas for shale gas production.

In view of the safe and efficient exploration and mining process of coal seams, Liu et al. investigated the physical similarity simulation of deformation and failure characteristics of coal–rock rise under the influence of repeated mining in close distance coal seams [15]. This research put forward an accurate laying of model and precise excavation of roadway test method which can effectively solve the problem of accurate laying of model and precise excavation of roadway in physical similarity simulation test of roadway with special surrounding rock structures.

Finally, as we stated above, due to the complex of the scientific questions and difficulties, there are still big challenges and several outstanding questions on the exploration, development and utilization of coal and coal-related resources, e.g., advanced extraction technologies of strategic metals from coal. Further studies are needed to fulfill the most efficient and clean utilization of coal and coal-related resources.

Author Contributions: Conceptualization, J.L., Y.C. and L.Z.; investigation, J.L.; resources, J.L., Y.C. and L.Z.; writing—original draft preparation, J.L.; writing—review and editing, Y.C. and L.Z.; funding acquisition, J.L. All authors have read and agreed to the published version of the manuscript.

Funding: This research was funded by the National Key Research and Development Program of China (No. 2021YFC2902000), and the National Natural Science Foundation of China (No. 41972179).

Acknowledgments: The editors would like to appreciate all of the authors for contributing their high-level research papers to this Special Issue. They have provided a wide variety of research objectives related to advances in exploration, development and utilization of coal and coal-related resources. We would like also to express our gratitude to several peer reviewers, who contributed a lot of constructive comments to the authors to improve the quality of the papers.

Conflicts of Interest: The authors declare no conflict of interest.

References

1. Li, J.; Zhuang, X.; Querol, X.; Font, O.; Moreno, N. A review on the applications of coal combustion products in China. *Int. Geol. Rev.* **2018**, *60*, 671–716. [CrossRef]
2. Dai, S.; Finkelman, R.B. Coal as a promising source of critical elements: Progress and future prospects. *Int. J. Coal Geol.* **2018**, *186*, 155–164. [CrossRef]
3. Dai Liu, C.; Zhao, L.; Liu, J.; Wang, X.; Ren, D. Strategic metal resources in coal-bearing strata; Significance and challenges. *J. China Coal Soc.* **2022**, *47*, 1743–1749. [CrossRef]
4. Ahmaruzzaman, M. A review on the utilization of fly ash. *Prog. Energy Combust. Sci.* **2010**, *36*, 327–363. [CrossRef]
5. Li, J.; Zhuang, X.; Font, O.; Moreno, N.; Vallejo, V.R.; Querol, X.; Tobias, A. Synthesis of merlinoite from Chinese coal fly ashes and its potential utilization as slow release K-fertilizer. *J. Hazard. Mater.* **2014**, *265*, 242–252. [CrossRef] [PubMed]
6. Li, J.; Zhuang, X.; Leiva, C.; Cornejo, A.; Font, O.; Querol, X.; Moreno, N.; Arenas, C.; Fernández-Pereira, C. Potential utilization of FGD gypsum and fly ash from a Chinese power plant for manufacturing fire-resistant panels. *Constr. Build. Mater.* **2015**, *95*, 910–921. [CrossRef]
7. Li, J.; Zhuang, X.; Monfort, E.; Querol, X.; Llaudis, A.S.; Font, O.; Moreno, N.; García Ten, F.J.; Izquierdo, M. Utilization of coal fly ash from a Chinese power plant for manufacturing highly insulating foam glass: Implications of physical, mechanical properties and environmental features. *Constr. Build. Mater.* **2018**, *175*, 64–76. [CrossRef]
8. Yao, Z.; Ji, X.; Sarker, P.; Tang, J.; Ge, L.; Xia, M.; Xi, Y. A comprehensive review on the applications of coal fly ash. *Earth Sci. Rev.* **2015**, *141*, 105–121. [CrossRef]
9. Dai, S.; Zheng, X.; Wang, X.; Finkelman, R.B.; Jiang, Y.; Ren, D.; Yan, X.; Zhou, Y. Stone coal in China: A review. *Int. Geol. Rev.* **2018**, *60*, 736–753. [CrossRef]
10. Li, B.; Zhang, F.; Liao, J.; Li, B.; Zhuang, X.; Querol, X.; Moreno, N.; Shangguan, Y. Geological controls on geochemical anomaly of the Carbonaceous mudstones in Xian'an Coalfield, Guangxi Province, China. *Energies* **2022**, *15*, 5196. [CrossRef]
11. Wei, Y.; He, W.; Qin, G.; Wang, A.; Cao, D. Mineralogy and geochemistry of the lower Cretaceous coals in the Junde Mine, Hegang Coalfield, Northeastern China. *Energies* **2022**, *15*, 5078. [CrossRef]
12. Zhang, L.; Zheng, L.; Liu, M. Study on the mineralogical and geochemical characteristics of arsenic in Permian coals: Focusing on the coalfields of Shanxi Formation in Northern China. *Energies* **2022**, *15*, 3185. [CrossRef]
13. Chen, W.; Wang, X.; Tu, M.; Qu, F.; Chao, W.; Chen, W.; Hou, S. The influence of fracturing fluid volume on the productivity of coalbed methane wells in the Southern Qinshui Basin. *Energies* **2022**, *15*, 7673. [CrossRef]
14. Xia, P.; Hao, F.; Tian, J.; Zhou, W.; Fu, Y.; Guo, C.; Yang, Z.; Li, K.; Wang, K. Depositional environment and organic matter enrichment of early Cambrian Niutitang black shales in the Upper Yangtze Region, China. *Energies* **2022**, *15*, 4551. [CrossRef]
15. Liu, P.; Gao, L.; Zhang, P.; Wu, G.; Wang, Y.; Liu, P.; Kang, X.; Ma, Z.; Kong, D.; Han, S. Physical similarity simulation of deformation and failure characteristics of coal-rock rise under the influence of repeated mining in close distance coal seams. *Energies* **2022**, *15*, 3503. [CrossRef]

Article

Geological Controls on Geochemical Anomaly of the Carbonaceous Mudstones in Xian'an Coalfield, Guangxi Province, China

Bo Li [1], Fuqiang Zhang [1,2], Jialong Liao [2], Baoqing Li [1,*], Xinguo Zhuang [1], Xavier Querol [1,3], Natalia Moreno [3] and Yunfei Shangguan [1]

[1] Key Laboratory of Tectonics and Petroleum Resources, China University of Geosciences, Ministry of Education, Wuhan 430074, China; boli@cug.edu.cn (B.L.); zhangfq@cug.edu.cn (F.Z.); xgzhuang@cug.edu.cn (X.Z.); xavier.querol@idaea.csic.es (X.Q.); sgyunfei@cug.edu.cn (Y.S.)
[2] Guangxi Bureau of Coal Geology, Liuzhou 545005, China; jialongliao01@gmail.com
[3] Institute of Environmental Assessment and Water Research, IDÆA-CSIC C/Jordi Girona, 18–26, 08034 Barcelona, Spain; natalia.moreno@idaea.csic.es
* Correspondence: libq@cug.edu.cn

Abstract: The anomalous enrichment of the rare earth elements and yttrium (REY), U, Mo, As, Se, and V in the coal-bearing intervals intercalated within the carbonate successions in South China has attracted much attention due to the highly promising recovery potential for these elements. This study investigates the mineralogical and geochemical characteristics of the late Permian coal-bearing intervals (layers A–F) intercalated in marine carbonate strata in the Xian'an Coalfield in Guangxi Province to elucidate the mode of occurrence and enrichment process of highly elevated elements. There are two mineralogical assemblages, including quartz-albite-kaolinite-carbonates assemblage in layers D–F and quartz-illite-kaolinite-carbonates assemblage in layers A–C. Compared to the upper continental crust composition (UCC), the REY, U, Mo, As, Se, and V are predominantly enriched in layers A and B, of which layer A displays the REY–V–Se–As assemblage while layer B shows the Mo–U–V assemblage. The elevated REY contents in layer B are primarily hosted by clay minerals, zircon, and monazite; Mo, U, and V show organic association; and As and Se primarily display Fe-sulfide association. Three geological factors are most likely responsible for geochemical anomaly: (1) the more intensive seawater invasion gives rise to higher sulfur, Co, Ni, As, and Se contents, as well as higher Sr/Ba ratio in layers A–C than in layers D–F; (2) both the input of alkaline pyroclastic materials and the solution/rock interaction jointly govern the anomalous enrichment of REY; and (3) the influx of syngenetic or early diagenetic hydrothermal fluids is the predominant source of U, Mo, V, Se, and As.

Keywords: geochemistry; rare earth elements and Y (REY); Yunkai Upland; Heshan Formation; mineral

---

Citation: Li, B.; Zhang, F.; Liao, J.; Li, B.; Zhuang, X.; Querol, X.; Moreno, N.; Shangguan, Y. Geological Controls on Geochemical Anomaly of the Carbonaceous Mudstones in Xian'an Coalfield, Guangxi Province, China. *Energies* **2022**, *15*, 5196. https://doi.org/10.3390/en15145196

Academic Editor: Reza Rezaee

Received: 28 June 2022
Accepted: 16 July 2022
Published: 18 July 2022

Publisher's Note: MDPI stays neutral with regard to jurisdictional claims in published maps and institutional affiliations.

1. Introduction

Coals formed in marine carbonate platforms are primarily distributed in South China, including Heshan [1–4], Fuisu [5], Xian'an [6], Yishan [7], Guiding [8], Yanshan (Yunnan Province) [9], and Chenxi (Hunan Province) coalfields [10], of which the first four are located within Guangxi Province, South China. These coals mostly contain highly enhanced concentrations of sulfur (especially organic sulfur) and, thus are classified as super-high-organic-sulfur (SHOS) coals [11,12].

The Late Permian is an important coal-forming period in Guangxi Province, South China. In recent years, the late Permian coal-bearing strata intercalated in marine carbonate strata in Guangxi Province, South China, have attracted much attention mainly due to the elevated concentrations of some critical elements, such as rare earth element and yttrium (REY), U, Mo, Se, V, and so on [3,5,6]. For example, the coals from the Heshan and Yishan

Coalfields are enriched in Mo, U, Se, and V [1–3,13] while those from the Fusui, Xian'an, and Yishan Coalfields are characterized by the elevated concentrations of REY, Zr (Hf), and Li [5,6,13].

The geological factors controlling the geochemical anomaly are still controversial. Dai et al. [3] attributed the elevated concentrations of Mo–U–Se–V assemblage in the coals from the Heshan and Yishan Coalfields in Guangxi Province to the joint influence of terrigenous detrital materials from Yunkai Upland and multistage low-temperature hydrothermal fluids. However, the marine invasion is also considered as being the predominant factor controlling the enrichment of Mo–U–Se–V assemblage in the coals from the Heshan Coalfield in Guangxi Province [1]. Additionally, Zeng et al. [13] attributed the enrichment of Mo, U, Se, and V in the coals from the Heshan Coalfield to the soil horizon at top of the middle Permian Maokou Formation. The anomalous enrichment of REY in the coals from the Yishan Coalfield is attributed to the influx of high-temperature hydrothermal fluids [13]. However, the joint influence of volcanic ash fall and water/rock interaction is regarded as a predominant factor influencing the enrichment of REY in the coals from the Xian'an Coalfield in Guangxi Province [6].

The Xian'an Coalfield contains relatively abundant coal resources, but the mineralogical and geochemical compositions of the coals are limitedly investigated [6] and report the enrichment of REY, Sc, U, Pb, and Mo in the lowermost coal seam of the Heshan Formation. However, whether the rare metals are enriched in other coal seams in the Xian'an Coalfield is still unclear. This study investigates the mineralogy and geochemistry of the late Permian coal-bearing intervals located in the middle and upper portions of the Heshan Formation in the Xian'an Coalfield, Guangxi Province, China, with an emphasis on the mode of occurrence and enrichment mechanism of some elements highly enriched in the coal-bearing intervals studied. It also provides an opportunity to determine whether the coal-bearing intervals in the Xian'an Coalfield can be considered as promising raw materials for certain rare metals.

2. Geological Setting

Late Permian is an important coal-forming period in South China, including Guizhou, Yunnan, Sichuan, and Guangxi Provinces [14]. The late Permian coalfields in Guangxi Province, South China, mainly include the Heshan, Fusui, Yishan, Xian'an, and Baiwang Coalfields (Figure 1A). The Shanglin exploration region studied is located within the Xian'an Coalfield in Guangxi Province.

The palaeogeographic environment of the late Permian in Guangxi Province is mainly represented by a series of isolated carbonate platforms surrounded by deep flume basins (Figure 1B). The late Permian strata include the Heshan and Dalong Formations, the former of which is approximately 140 m thick and consists mainly of carbonates intercalated with coal seam, mudstone, and carbonaceous mudstone while the latter of which is mainly composed of siliceous rocks and siltstones intercalated with tuffaceous sandstone and marlstone.

The late Permian coal-bearing stratum in the Xian'an Coalfield is primarily the Heshan Formation. There are mainly five layers of coal in the Heshan Formation, which are numbered from top to bottom as K_1, K_2, K_3, K_4, and K_5 coals, respectively (Figure 2). The K_3 and K_4 coals are divided into two or three layers in this area. The lithology of the roof and floor of the coals is limestone or flint (Figure 2). Based on the lithological and coal-bearing characteristics, the Heshan Formation is subdivided into the upper and lower sections (Figure 2). The upper section of the Heshan Formation is the primary coal-bearing stratum and consists of bioclastic limestone and four coal seams while the lower section is a secondary coal-bearing stratum and composed of limestone, biolimestone, and gravel clastic limestone intercalated with one locally mined coal seam K_5. The total thickness of the coal seam is 11.6 m, and the total recoverable thickness is 6.4 m, among which the K_4 coal seam is the main mineable coal seam with an average of 1.80 m, and the K_3 and K_2 coal seams are locally workable or unworkable [1].

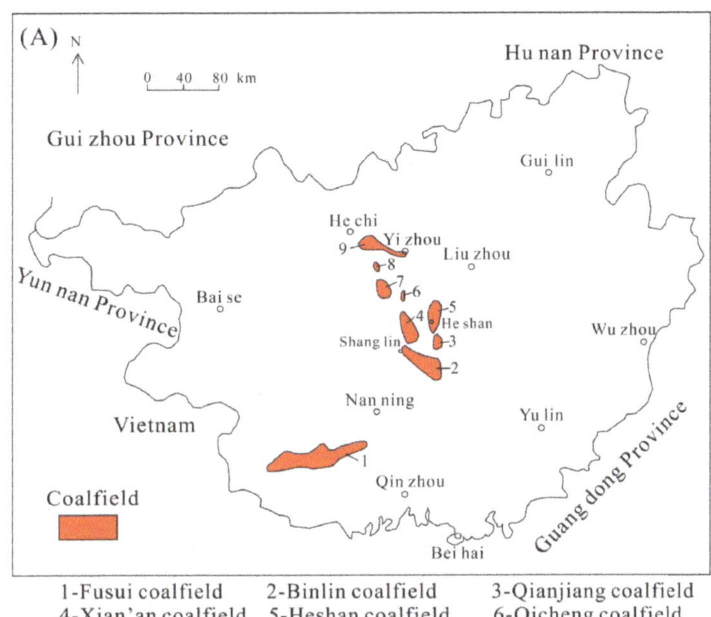

1-Fusui coalfield   2-Binlin coalfield   3-Qianjiang coalfield
4-Xian'an coalfield   5-Heshan coalfield   6-Qicheng coalfield
7-Baiwang coalfield   8-Duanlaren coalfield   9-Yishan coalfield

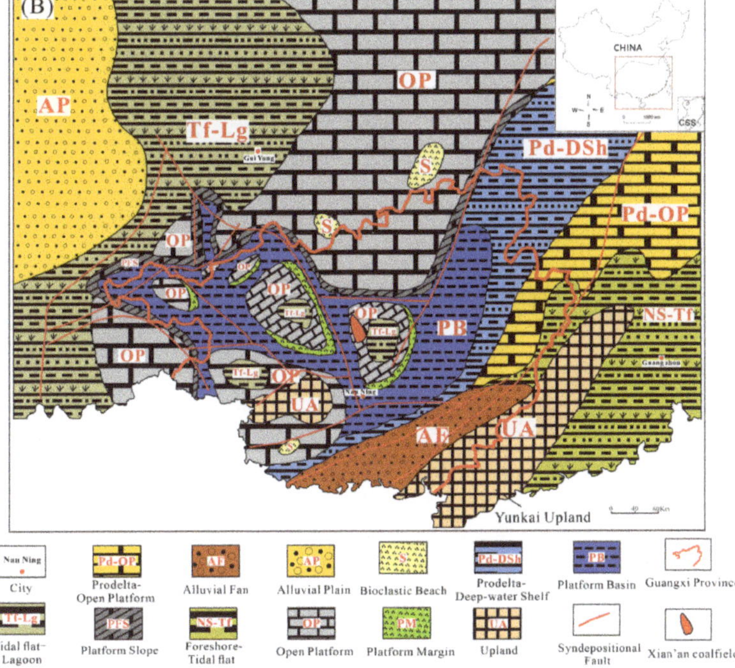

Figure 1. (**A**) Location of coalfields in Guangxi Province [15]; (**B**) location of the Xian'an Coalfield as well as the distribution of Late Permian sedimentary environments in southern China. CSS, China South Sea [16].

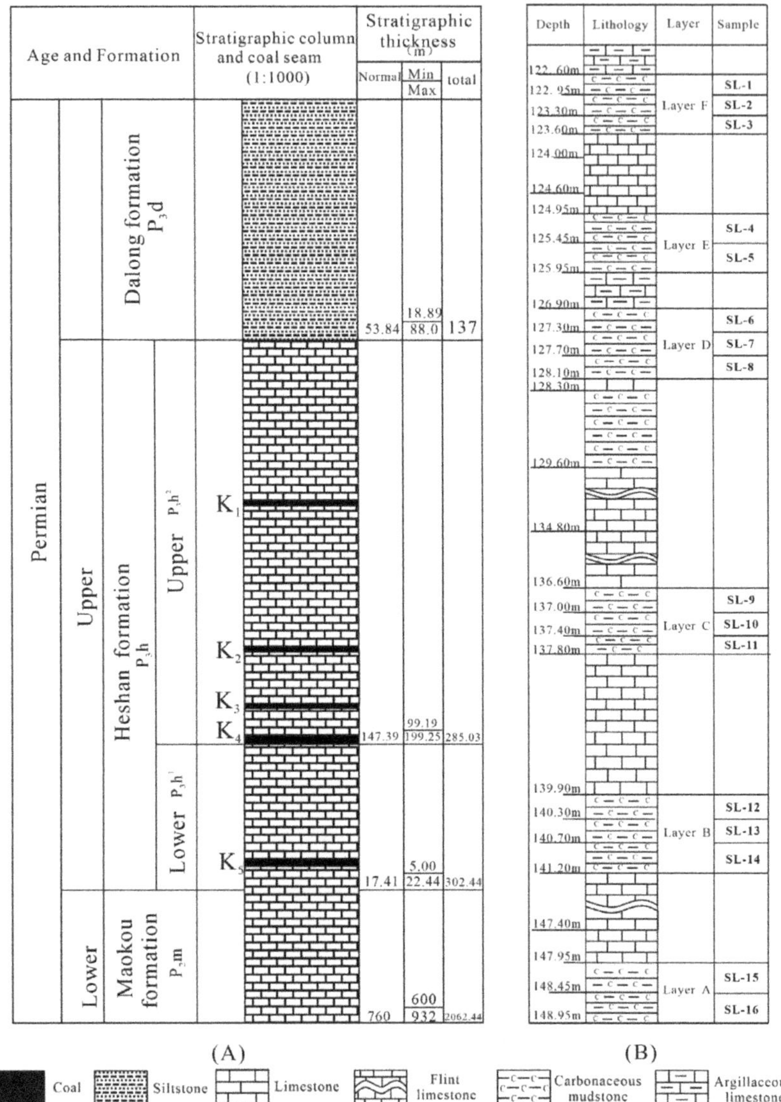

Figure 2. (**A**) The generalized stratigraphic column of the late Permian coal-bearing strata within the Xian'an Coalfield [6]; (**B**) sampling column of borehole SL.

3. Methodology

Sixteen samples of carbonaceous mudstone were collected from borehole SL located in the Wanfu exploration region of the Xian'an Coalfield in Guangxi Province, SW China (Figure 2B). The samples collected were then put in plastic bags to avoid contamination and oxidation.

The individual samples were ground to ≤0.2 mm and split into two representative portions. A portion of the sample (<0.2 mm) was directly used for proximate analysis based on the ASTM Standards D3173-11 (2011), D3174-11 (2011), and D3175-11 (2011) [17–19], while another portion was ground further to ≤0.076 mm (200 mesh) using an agate mortar and pestle for mineralogical and geochemical analysis. Mineralogical analyses of the sample powders were performed by powder X-ray Diffraction (XRD) using a Bruker D8 A25 Advance (Bruker D8 A25 Advance, Leipzig, Germany) at Institute of Environmental Assessment and Water Research (Barcelona, Spain). The detailed XRD analysis procedure and semi-quantitative analysis were reported in the previous study [20]. The diffractograms were obtained at a 2θ interval of 5–90°, with a step size of 0.01°.

Approximately, a 0.1 g sample was weighed and digested based on the method proposed by Querol et al. [21] for geochemical analysis. Major elements (Al, Ti, Fe, Mg, Ca, Na, K, and P) and trace elements were performed by inductively coupled plasma atomic-emission spectrometry (ICP-AES, Iris Advantage TJA Solutions, Thermo Fisher Scientific, Waltham, MA, USA) and inductively coupled plasma mass spectrometry (ICP-MS, X-Series II Thermo, Thermo Fisher Scientific, Waltham, MA, USA). Silicon content was measured by wavelength dispersive X-ray fluorescence spectrometry (XRF; ZSXPrimus II) following the methods for chemical analysis of silicate rocks (GB/T14506.28-2010).

A small portion of representative block samples was used to prepare the polished sections for the SEM-EDS analysis. The modes of occurrence of minerals were studied using field emission-scanning electron microscope (ZEISS Sigma300, Carl Zeiss AG, Jena, Germany), equipped with an energy-dispersive X-ray spectrometer (EDS) in the State Key Laboratory of Geological Processes and Mineral Resources (China).

4. Results

4.1. Coal Chemistry

The moisture content, ash yield, volatile matter yield, and total sulfur content of the samples from borehole SL are tabulated in Table 1. The moisture content and volatile matter yield of the samples range from 0.4% to 2.1% and 8.4% to 16.3%, respectively. The samples are characterized by high ash yield, which ranges from 54.4% to 86.9% with an average of 75.8% exceeding 50%, and thus are classified as carbonaceous mudstone rather than coal according to Chinese standard (>50% ash yield indicative of noncoal rock, GB/T 15224.1-2018). However, these carbonaceous mudstones are used as high-ash coals due to relatively rare coal resources in Guangxi Province. Vertically, the ash yield is distinctly lower in layers A and C than in the other layers (Figure 3). The low-temperature ash yield (LTA) of the investigated samples is higher than the high-temperature ash yield (Table 1). The difference is partly due to dehydration of the clay minerals, oxidation of the pyrite, and/or CO_2 release from the carbonate minerals during the high-temperature ashing process [3].

Table 1. Proximate analysis, total sulfur content, and low-temperature ash yield (LTA) of the samples from borehole SL.

Layer	Sample	LTA (d, wt%)	Moisture (ad, wt%)	Ash Yield (d, wt%)	Volatile Matter Yield (daf, wt%)	Total Sulfur Content (d, wt%)
Layer F	SL-1	93.2	0.39	84.8	12.0	2.1
	SL-2	93.0	0.73	84.1	10.0	2.7
	SL-3	96.3	1.17	83.2	10.7	2.4
Layer E	SL-4	96.1	0.58	86.3	10.2	2.5
	SL-5	88.9	0.80	86.9	10.6	2.0
Layer D	SL-6	89.7	0.59	82.5	9.6	2.1
	SL-7	77.6	0.37	69.0	10.1	4.6
	SL-8	88.7	0.79	80.0	9.8	2.7
Layer C	SL-9	85.6	1.32	70.0	13.2	8.6
	SL-10	72.0	1.33	59.2	10.8	5.1
	SL-11	83.6	1.59	63.3	14.3	6.9
Layer B	SL-12	93.8	2.14	86.2	12.2	2.8
	SL-13	97.5	2.00	83.2	16.1	6.3
	SL-14	89.7	1.18	72.3	15.4	4.9
Layer A	SL-15	78.7	0.60	54.4	9.8	7.7
	SL-16	83.0	0.60	65.4	8.4	4.0

Note: ad, air-dry basis; d, dry basis; daf, dry and ash-free basis.

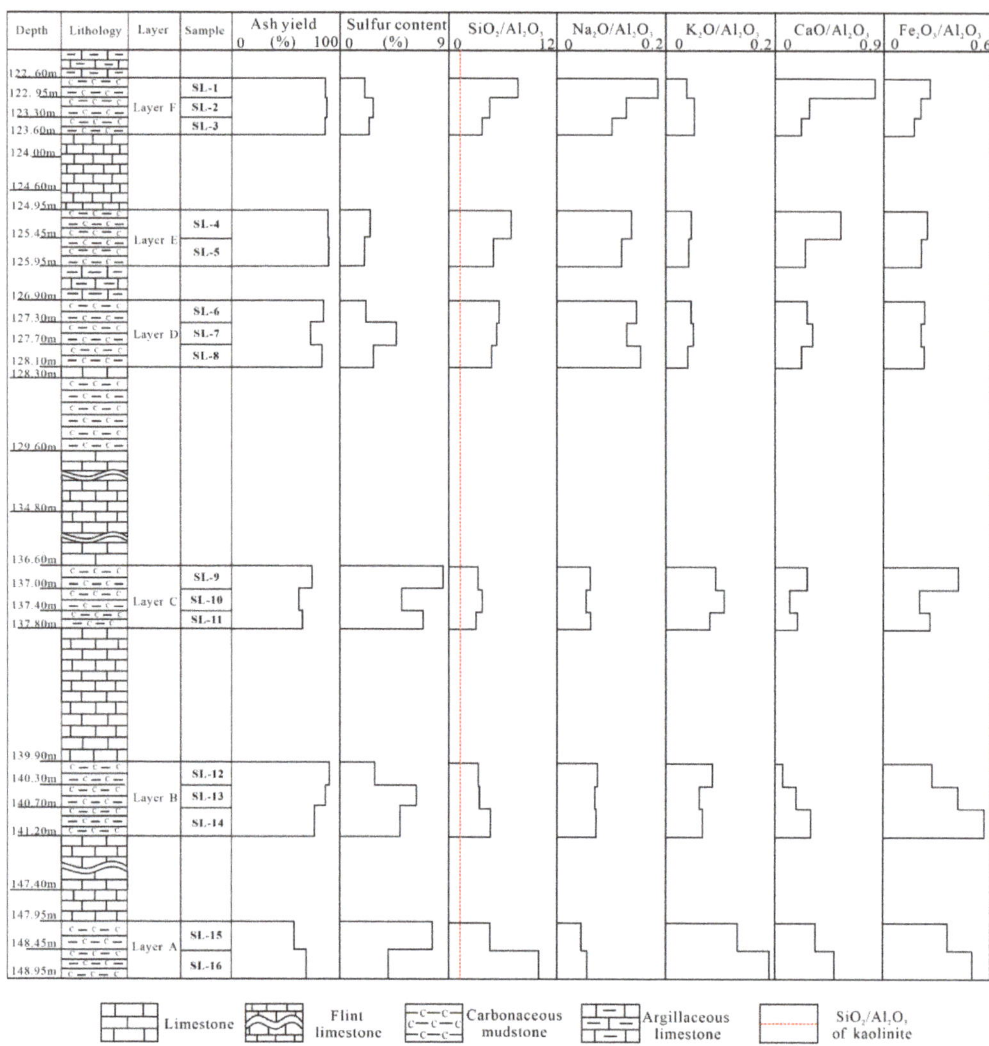

Figure 3. Vertical distribution of ash yield, sulfur content, and the ratios of major element oxides throughout borehole SL profile.

The total sulfur contents of the samples from borehole SL vary between 2.0% and 8.6%, with an average of 4.2%, indicating a high sulfur content for them (<1.0, 1.0–3.0, and >3.0% indicative of low, medium, and high sulfur content, respectively) [12]. The sulfur content shows a distinct variation throughout the profile and displays higher sulfur contents in layers A–C than in layers D–F (Figure 3). Layer A (5.8% on average), B (4.7%), C (6.8%), and D (3.2%) contain a high sulfur content (>3.0%) while layers E (2.2%) and F (2.4%) are characterized by medium sulfur content.

4.2. Mineralogy

4.2.1. Mineral Phases

The contents of the crystalline mineral phases of the samples taken from borehole SL are tabulated in Table 2. The minerals in the investigated samples consist mainly of quartz, and to a lesser extent, illite, albite, calcite, dolomite, kaolinite, and pyrite, along with trace amounts of bassanite and anatase (Figure 4). Paragonite is only present in sample SL-5. There are two types of mineral assemblage throughout the borehole SL profile. The first mineral assemblage (pyrite-quartz-albite-kaolinite-carbonates) is only present in layers D–F (Figure 5) while the second assemblage (pyrite-quartz-illite-kaolinite-carbonates) occurs in layers A–C (Figure 5).

Table 2. Mineralogical proportions and low-temperature ash yield (LTA) of the samples from the borehole SL (on whole-coal basis; unit in wt%).

Sample	LTA	Illite	Kaolinite	Paragonite	Quartz	Calcite	Dolomite	Pyrite	Albite	Anatase	Bassanite
SL-1	93.2	<dL	5.6	<dL	55.2	8.4	8.6	2.1	13.4	<dl	<dL
SL-2	93.0	<dL	7.1	<dL	48.3	1.1	4.9	2.0	29.6	<dl	<dL
SL-3	96.3	<dL	6.9	<dL	58.8	1.6	14.0	2.2	12.9	<dl	<dL
SL-4	96.1	<dL	3.2	<dL	60.6	2.5	10.2	2.0	17.1	<dl	0.4
SL-5	88.9	<dL	5.7	5.3	55.1	0.7	9.7	2.7	9.6	0.1	<dL
SL-6	89.7	<dL	<dl	<dL	59.8	<dL	9.5	4.1	16.3	<dl	<dL
SL-7	77.6	<dL	6.2	<dL	54.2	<dL	4.6	2.9	9.7	<dl	<dL
SL-8	88.7	<dL	2.8	<dL	55.0	1.8	8.9	3.8	16.5	<dl	<dL
SL-9	85.6	12.3	6.4	<dL	45.3	5.8	<dL	7.5	<dL	0.3	8.0
SL-10	72.0	15.0	15.1	<dL	33.7	2.0	2.5	3.7	<dL	<dl	<dL
SL-11	83.6	14.8	3.7	<dL	47.6	3.0	4.0	10.5	<dL	<dl	<dL
SL-12	93.8	17.6	<dl	<dL	64.8	<dL	4.4	6.5	<dL	0.4	<dL
SL-13	97.5	12.5	9.4	<dL	52.3	<dL	12.2	7.4	<dL	0.3	3.4
SL-14	89.7	7.8	7.5	<dL	62.4	<dL	2.7	9.3	<dL	<dL	<dL
SL-15	78.7	11.5	2.1	<dL	48.1	7.3	3.9	5.9	<dL	<dL	<dL
SL-16	83.0	11.8	1.3	<dL	59.3	6.2	0.5	3.9	<dL	<dL	<dL

<dL, below detection limit.

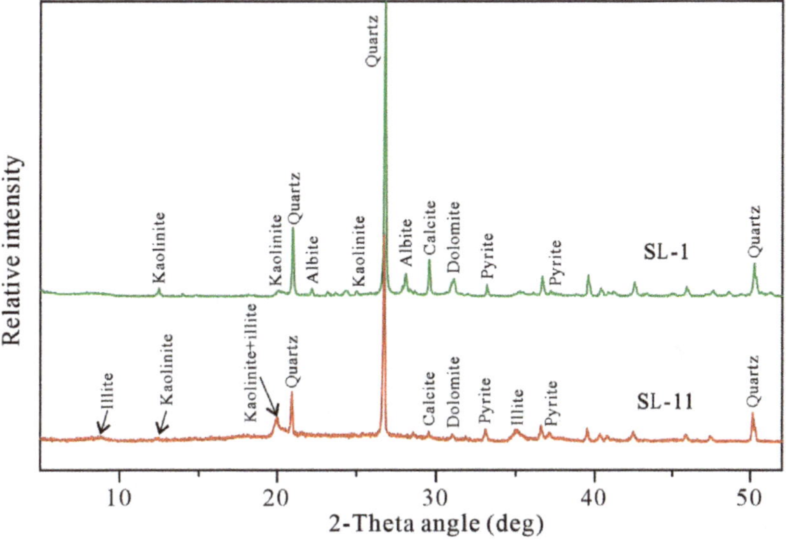

Figure 4. X-ray diffractogram (XRD) patterns of minerals in selected samples to show mineral assemblage.

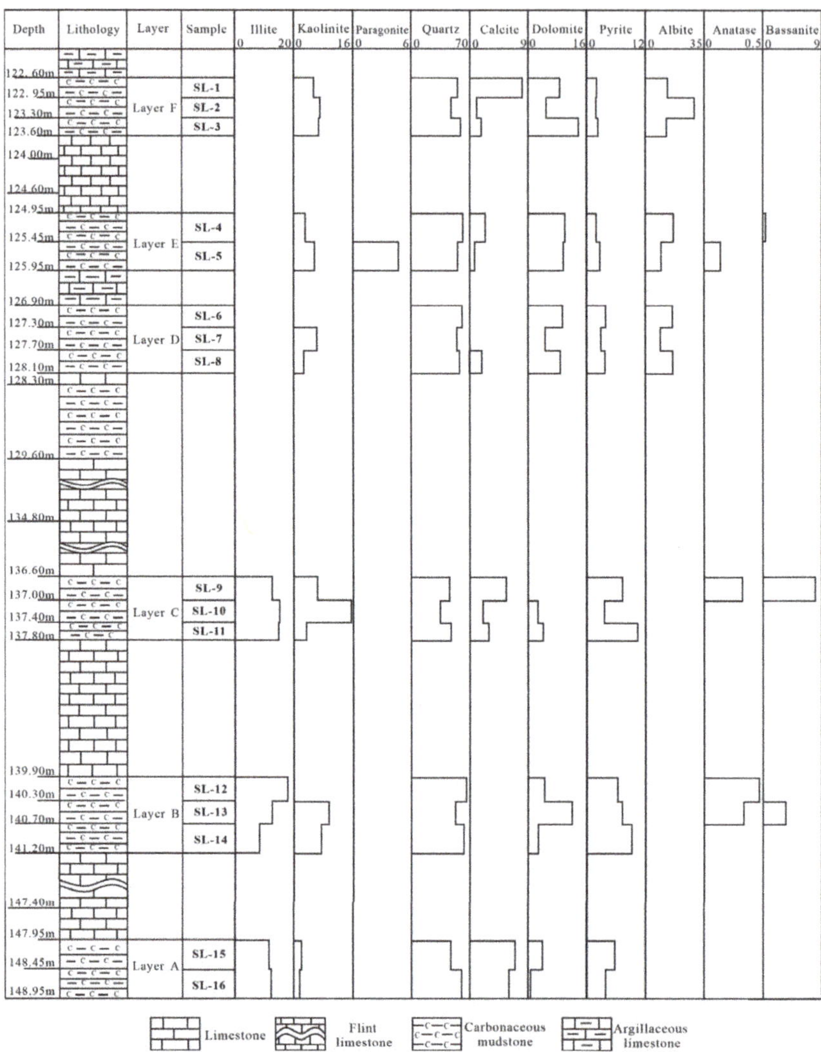

Figure 5. Vertical distribution of minerals throughout borehole SL profile.

Illite is only distributed in layer A (11.7% on average), layer B (12.6%), and layer C (14.0%) while kaolinite universally occurs in all layers (Figure 5). Similarly, pyrite is also abundant in layers A–C relative to layers D–F. Dolomite is more abundant in layers B, D, E, and F compared to layers A and C. The distribution of calcite, however, is different from that of dolomite. Notably, albite is a major mineral constituent in layers D–F but is absent in layers A–C (Figure 5).

4.2.2. Mode of Occurrence of Minerals

Kaolinite mainly occurs in the following forms: kaolinite matrix (Figure 6A), pore/cavity-filling kaolinite (Figure 6B), vermiculate kaolinite (Figure 6C), and fracture-filling kaolinite (Figure 6D), of which the former three forms indicate the syngenetic to early diagenetic stage while the fourth suggests epigenetic stage.

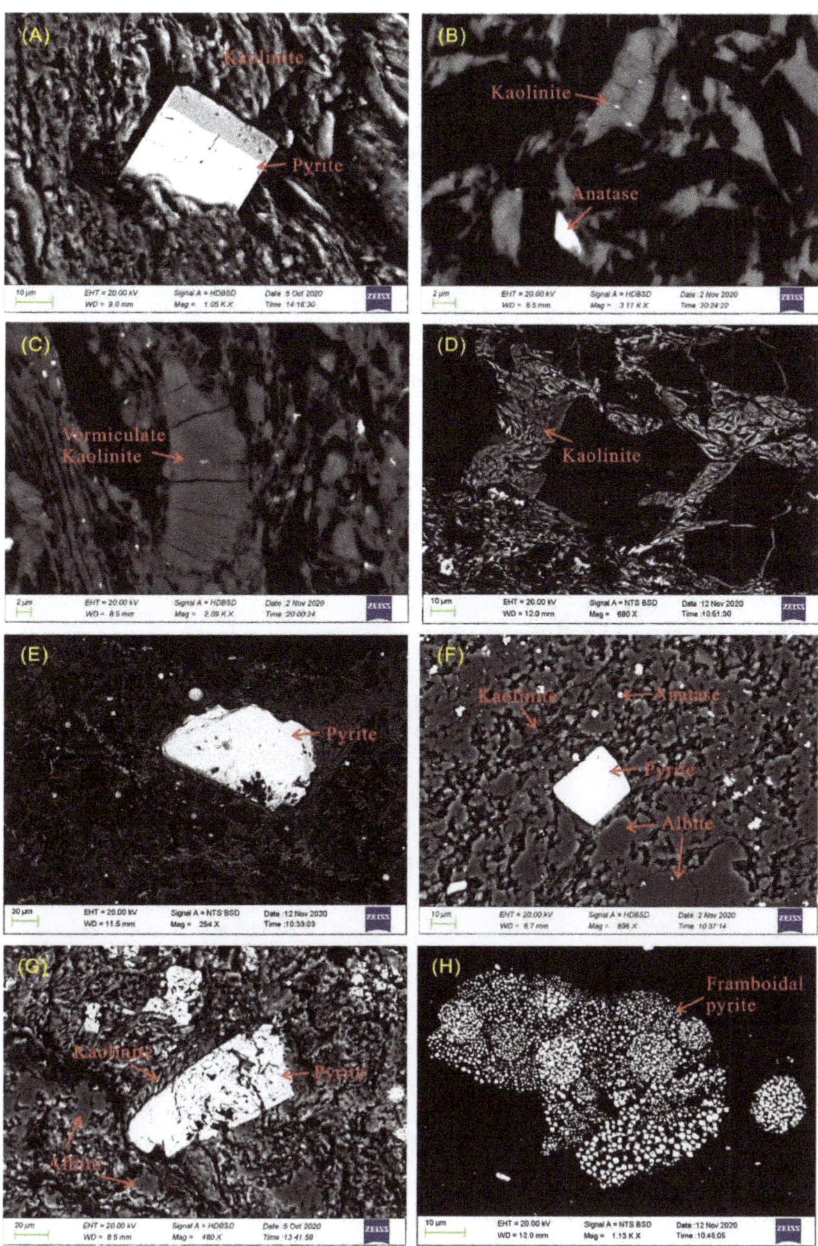

Figure 6. Scanning electron microscope (SEM) back-scattered electron images of minerals (**A–H**): (**A**) euhedral pyrite embedded within kaolinite matrix (sample SL-9); (**B**) kaolinite and anatase (sample SL-9); (**C**) vermiculate kaolinite (sample SL-12); (**D**) fracture-filling kaolinite (sample SL-5); (**E**) massive pyrite (sample SL-5); (**F**) albite, kaolinite, pyrite, and finely-grained disseminated anatase particles (sample SL-5); (**G**) kaolinite, albite, and pyrite (sample SL-5); and (**H**) framboidal pyrite aggregates (sample SL-16).

Pyrite is a primary mineral phase in samples studied. Pyrite primarily occurs as single euhedral crystals (Figure 6A) or massive forms (Figure 6E–G) embedded in the kaolinite matrix, which indicates an approximately contemporaneous early diagenetic formation. In a few cases, framboidal pyrite aggregates are also observed in the samples studied (Figure 6H).

Albite is an important mineral constituent in layers D–F. Albite predominantly occurs as disseminated particles with irregular corroded borders (Figure 6F,G), indicating the alteration of albite. In most cases, albite is surrounded by flaky kaolinite, revealing the transformation of albite to kaolinite.

Anatase is also found in some samples and primarily occurs as dispersed particles embedded within a clay minerals matrix (Figure 6F), indicating a detrital origin. In a few cases, anatase is embedded within organic matter (Figure 6B), indicating an authigenic origin.

4.3. Geochemistry

The major and trace elements contents of the samples collected from borehole SL are listed in Table 3.

Table 3. Major-element oxides (wt%) and trace element concentrations (μg/g) of the samples from borehole SL-1 (on whole-coal basis).

	SL-1	SL-2	SL-3	SL-4	SL-5	SL-6	SL-7	SL-8	SL-9	SL-10	SL-11	SL-12	SL-13	SL-14	SL-15	SL-16
SiO_2	66	70	69	72	67	69	62	66	55	48	57	73	63	70	55	66
TiO_2	0.17	0.23	0.36	0.20	0.22	0.18	0.13	0.25	0.54	0.34	0.30	0.51	0.41	0.26	0.31	0.21
Al_2O_3	7.2	12	13	8.3	12	10	9.1	12	12	11	13	17	15	9.7	7.3	4.4
Fe_2O_3	1.8	2.5	2.3	1.9	2.4	2.2	1.8	2.7	5.1	2.3	3.5	5.1	6.2	5.6	2.6	2.1
MgO	1.8	1.8	2.0	2.6	2.6	2.3	2.0	2.1	0.67	0.63	0.75	0.76	2.0	1.9	0.87	0.32
CaO	6.0	3.5	3.0	4.6	3.2	2.7	2.8	2.7	3.3	1.4	2.5	1.1	2.9	3.0	2.5	2.2
K_2O	0.29	0.62	0.74	0.38	0.51	0.47	0.45	0.51	1.1	1.2	1.0	1.4	0.91	0.62	0.95	0.85
Na_2O	1.4	1.6	1.5	1.2	1.5	1.5	1.2	1.9	0.73	0.60	0.84	1.3	1.1	0.71	0.31	0.26
Li	16	22	23	28	55	35	32	45	40	35	55	27	25	23	14	8.0
Be	<dl	1.7	1.4	1.6	1.8	0.98	2.1	0.85	1.4	1.2	1.6	3.6	3.1	0.84	<dl	1.0
B	108	241	115	77	81	98	65	46	82	96	166	98	134	66	79	45
P	98	97	219	82	100	113	97	222	168	69	81	372	139	77	62	635
Sc	4.1	5.4	9.7	5.3	7.6	6.6	5.9	7.8	6.9	11	8.5	14	12	7.7	8.2	3.7
V	71	70	54	56	57	120	216	126	75	137	100	424	248	359	504	424
Cr	28	14	15	87	32	73	100	75	33	51	28	298	121	444	357	165
Mn	95	102	138	129	136	160	132	168	405	122	208	49	83	93	85	75
Co	5.6	4.5	3.4	2.8	3.7	3.0	2.6	3.5	9.0	3.7	4.5	11	12	8.0	6.2	5.3
Ni	18	11	6.8	31	14	22	22	23	19	21	13	44	49	77	84	92
Cu	20	10	27	17	10	32	16	23	23	21	15	27	23	26	23	23
Zn	62	41	59	45	50	71	62	66	95	83	74	80	64	62	66	72
Ga	9.0	14	18	11	17	15	15	18	19	22	20	36	27	15	11	4.0
Ge	1.2	1.3	1.4	1.1	1.5	1.7	1.6	1.8	1.5	1.8	1.8	3.4	2.5	1.7	1.1	<dl
As	7.6	10	11	9.1	15	12	9.4	13	21	9.2	15	26	48	40	17	14
Se	3.0	2.7	4.4	2.3	4.4	4.3	4.0	5.5	3.4	4.0	5.4	15	10.0	7.8	5.8	5.2
Rb	16	35	44	22	30	27	24	26	42	49	42	71	46	30	39	20
Sr	495	664	650	525	567	554	514	567	677	501	771	798	825	539	508	337
Y	19	23	28	19	23	30	49	35	11	33	33	38	33	53	43	29
Zr	67	103	186	89	108	124	132	148	54	124	122	184	167	137	146	63
Nb	6.5	7.0	14	4.8	7.3	12	19	14	5.3	12	8.6	19	15	10	9.9	1.9
Mo	20	25	13	22	31	36	69	43	38	39	35	8.7	45	101	149	126
Cs	4.3	9.7	12	6.2	8.9	9.7	7.7	8.5	15	17	16	27	19	13	9.4	6.7
Ba	84	221	250	132	178	170	140	239	74	91	81	107	81	51	74	45
La	22	18	28	21	26	29	30	38	17	32	34	99	105	44	21	11
Ce	36	32	49	35	45	50	51	66	32	59	62	313	208	86	41	23
Pr	4.6	4.1	6.2	4.3	5.8	6.4	6.4	8.2	4.3	7.4	7.8	24	21	9.8	5.5	3.7
Nd	18	15	23	16	22	24	23	30	17	27	30	88	70	36	22	15
Sm	4.1	4.1	6.0	3.9	6.4	5.6	5.4	7.3	3.8	6.5	6.5	16	14	8.7	5.3	4.3
Eu	<dl	<dl	1.0	<dl	0.90	0.83	<dl	1.0	<dl	<dl	1.1	2.5	1.6	1.2	0.85	<dl
Gd	2.8	3.0	4.2	2.8	4.3	4.4	4.2	4.7	2.0	4.5	3.8	12	8.0	6.5	4.0	3.2
Tb	<dl	<dl	0.82	<dl	0.83	0.88	0.99	0.95	<dl	0.92	0.72	2.0	1.4	1.5	0.92	<dl
Dy	3.7	4.5	5.7	3.6	5.0	6.1	8.3	6.8	2.3	6.3	5.3	11	8.7	11	6.4	5.2
Ho	0.83	1.0	1.3	<dl	1.1	1.3	2.1	1.6	<dl	1.4	1.3	2.2	1.8	2.6	1.6	1.3
Er	1.9	2.4	3.0	1.8	2.1	2.9	5.5	3.8	1.1	3.1	2.7	4.5	4.0	6.0	3.7	2.8
Tm	<dl	0.80	0.49	<dl	0.36	0.51	2.0	0.62	<dl	1.0	0.42	0.75	0.68	1.0	0.58	0.95
Yb	2.4	3.2	4.2	2.1	2.6	4.0	8.3	5.2	1.3	4.0	3.2	6.3	5.5	8.1	4.5	3.8
Lu	<dl	<dl	0.69	<dl	0.44	0.66	1.4	0.87	<dl	<dl	0.54	1.0	0.92	1.3	0.76	<dl
Hf	1.8	3.0	4.7	2.6	3.6	3.4	3.1	3.9	1.8	3.8	3.6	5.4	4.8	3.5	3.4	1.4
Ta	<dl	<dl	<dl	<dl	<dl	<dl	<dl	<dl	<dl	<dl	<dl	<dl	<dl	<dl	<dl	<dl
Pb	19	32	29	17	26	26	26	29	25	34	32	66	38	35	31	32
Th	7.3	17	15	9.6	16	15	15	20	7.8	15	11	21	19	13	7.8	5.2
U	20	23	19	11	14	31	85	43	19	42	33	44	25	60	126	125

<dl, below detection limit.

4.3.1. Major Elements

The major elements in the investigated samples are predominantly composed of SiO_2, and, to a lesser extent, Al_2O_3, with the remaining Fe_2O_3, CaO, MgO, Na_2O, and K_2O as minor or trace components. The SiO_2/Al_2O_3 ratio (4.2–15.1, 6.6 on average) in the studied samples is evidently higher than the theoretical value of kaolinite (1.2) (Figure 3), illite (3.2), and albite (3.3), which would be explained by quartz-dominated mineral assemblage. The Na_2O/Al_2O_3 value is obviously higher in layers D–F than in layers A–C (Figure 3), coinciding with the mineral assemblage where albite is an abundant constituent in layers D–F but is absent in layers A–C as mentioned above. The higher K_2O/Al_2O_3 and Fe_2O_3/Al_2O_3 ratios in layers A–C than in layers D–F (Figure 3) would be ascribed to much more abundant occurrence of illite and pyrite, respectively, in layers A–C than layers D–F (Figure 5). Compared to layers A–C, layers D–F exhibit a distinctly higher CaO/Al_2O_3 ratio (Figure 3), which is due to the relatively abundant occurrence of carbonate minerals in layers D–F. This is also attested by the relatively significant correlation between MgO and dolomite (Figure 7A).

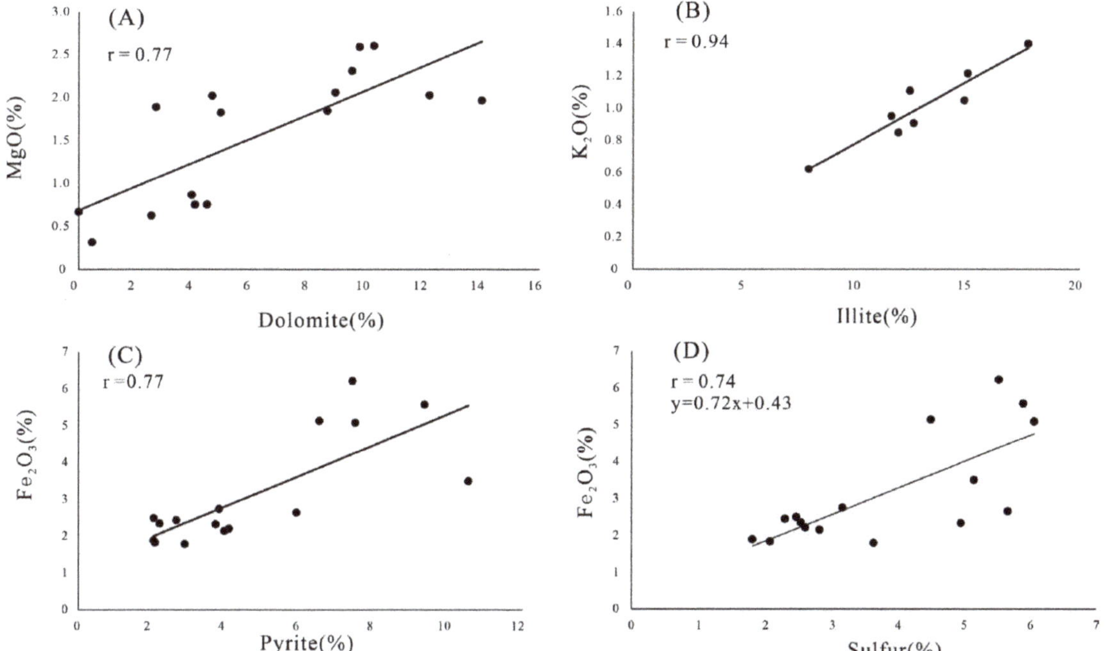

Figure 7. Relationship among selected elements and minerals: (**A**) plot of MgO vs. dolomite; (**B**) plot of K_2O vs. illite; (**C**) plot of Fe_2O_3 vs. pyrite; and (**D**) plot of Fe_2O_3 vs. sulfur.

K_2O significantly positively correlates with illite in layers A–C (Figure 7B), indicating an illite affinity of K_2O. Fe_2O_3 and pyrite have a relatively significant linear correlation (Figure 7C), suggesting a sulfide affinity for Fe. Additionally, Fe and sulfur display a relatively significant correlation (Figure 7D); however, the slope (0.72) of the Fe–S regression equation is distinctly lower than the theoretical Fe/S ratio (0.87), appearing to indicate that other forms of sulfur (e.g., organic sulfur) contribute to the total sulfur.

4.3.2. Trace Elements

To better elucidate the degree of enrichment or depletion of trace elements in coal-bearing strata, the concentration coefficient (CC) is used in the present study; the CC is a ratio of the studied samples versus referenced rocks, such as upper continental crust composition (UCC) [22] and world coal [23], with CC < 0.5, 0.5–2, 2–5, 5–10, and >10 indicative of depletion, similarity, slight enrichment, enrichment, and significant enrichment, respectively [8]. Compared to the average of trace elements in UCC [22], Se, Mo, and U are significantly enriched in the studied samples; As and B are enriched; V, Cr, Cs, Dy, Ho, Tm, and Yb are slightly enriched; and the remaining trace elements are depletion or similar to the values of UCC (Figure 8).

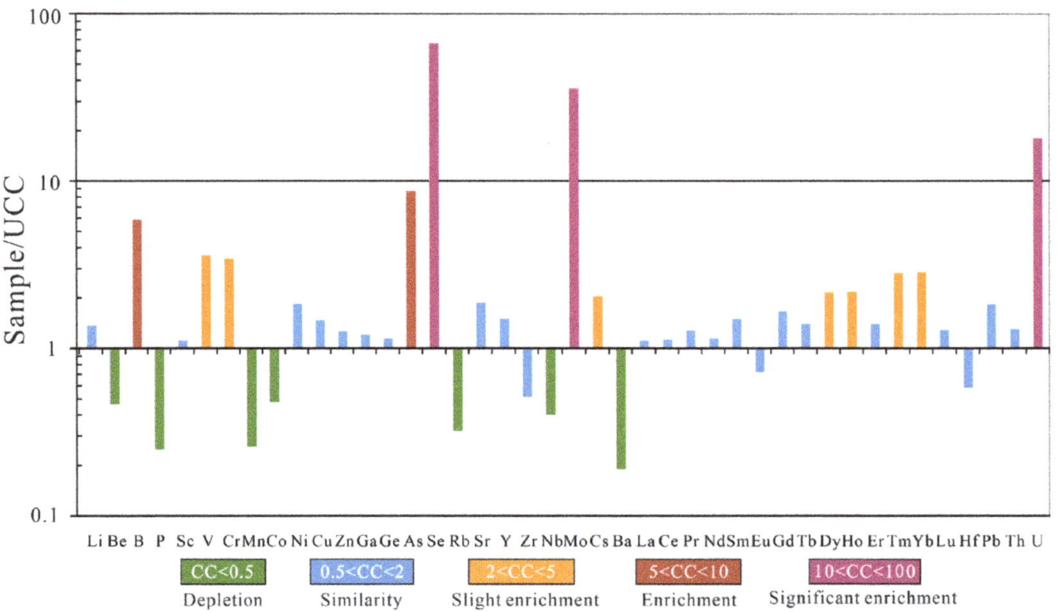

Figure 8. Concentration coefficient (CC) of trace elements in the studied samples, normalized to the respective average of upper continental crust composition (UCC; [22]).

Elements U, Mo, and Se in abundance show distinct variations throughout the vertical profile (Figure 9). The significantly enriched-U intervals are distributed in layer A (CC = 41.7), layer B (13.0), layer C (12.5), and layer D (CC = 21.2); enriched-U intervals are vertically located within layer E (CC = 5.1) and layer F (CC = 8.2). Mo and Se are significantly enriched among layers A–F, but they appear to be more elevated in layers A–D relative to layers E–F. As is significantly concentrated in layers A (CC = 11.9) and B (12.0), and other layers reach the enriched or slightly enriched level. Boron in abundance is enriched in layers B (CC = 6.8), C (CC = 6.8), and F (CC = 9.1) and slightly elevated in other horizons. Elements V, Cr, Dy, Ho, Tm, and Yb contents reach the enriched or slightly enriched level within layers A, B, and D.

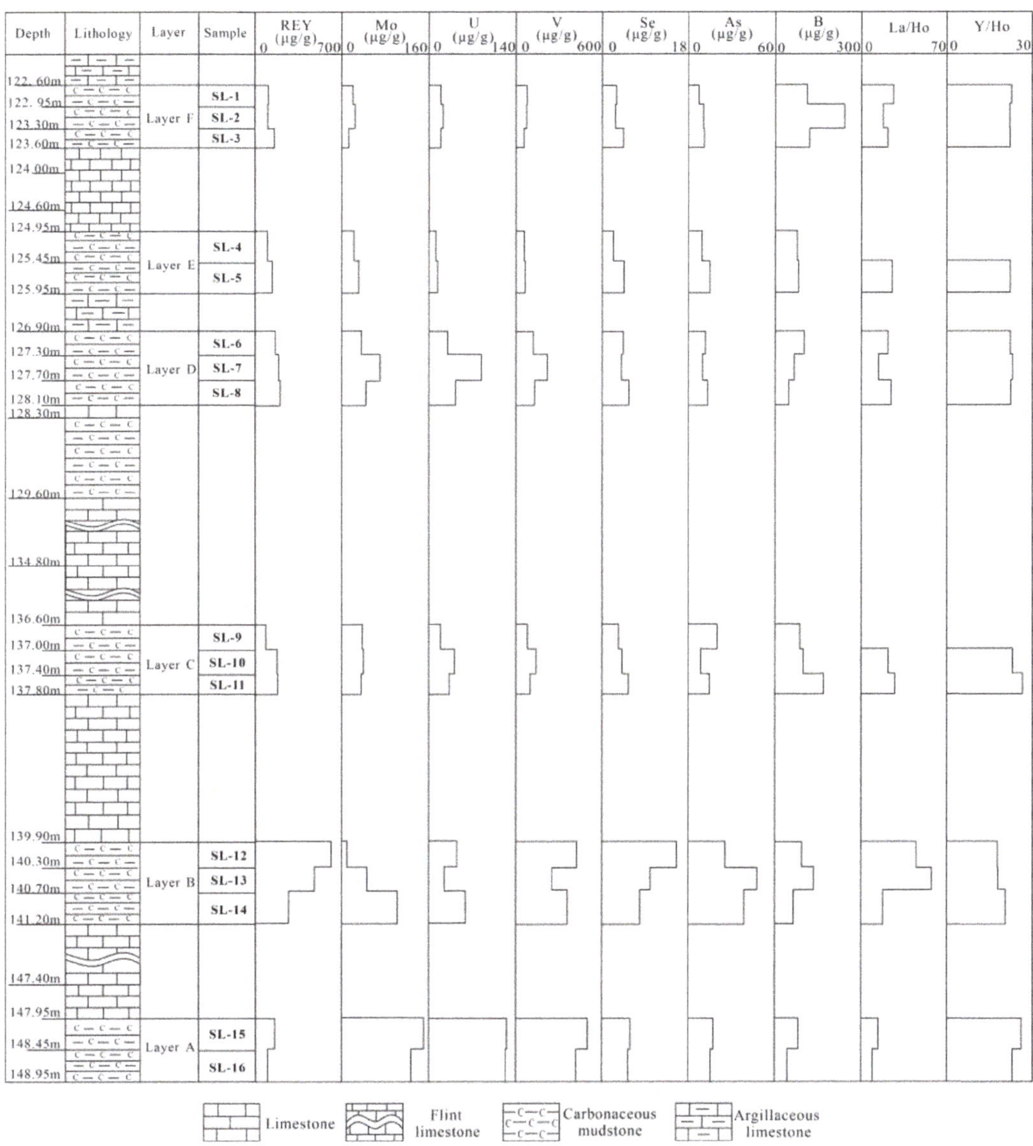

Figure 9. Vertical distribution of selected trace elements, La/Ho and Y/Ho, throughout borehole SL profile.

4.3.3. Rare Earth Element and Yttrium (REY)

The rare earth elements and yttrium (REY) contents in the investigated samples range from 92 µg/g to 625 µg/g (625 µg/g; sample SL-12) with an average of 208 µg/g, which is similar to that of UCC (168 µg/g; [22]) but higher than that of world coal (68.5 µg/g; [23]).

A three-fold REY classification, namely LREY (La, Ce, Pr, Nd, and Sm), MREY (Eu, Gd, Tb, Dy, and Y), and HREY (Ho, Er, Tm, Yb, and Lu), is used in the present study. The UCC-normalized REY distribution pattern is used in the present study to elucidate the distribution and fractionation of REY. The REY distribution pattern can be represented by three types, namely LREE distribution type (L-type), MREY distribution type (M-type), and HREY distribution type (H-type) [24]. The UCC-normalized REY distribution of the investigated samples shows two patterns of REY distribution (Figure 10). The first pattern is represented by the HREY distribution type (H-type) in the samples containing REY content similar to that of UCC (Figure 10A). The second pattern is represented by the LREY distribution type (L-type) in the samples (e.g., SL-12, SL-13) with elevated concentrations of REY (Figure 10B). Moreover, all samples are universally characterized by weakly negative Eu anomaly (0.67–0.95, 0.77 on average) and slightly negative Y anomaly (0.63–0.95, 0.80 on average) (Figure 10). Except for samples SL-12 and SL-13 showing a pronounced positive Ce anomaly and no Ce anomaly, respectively, other samples display slightly negative Ce anomaly (0.82–0.95, 0.85 on average).

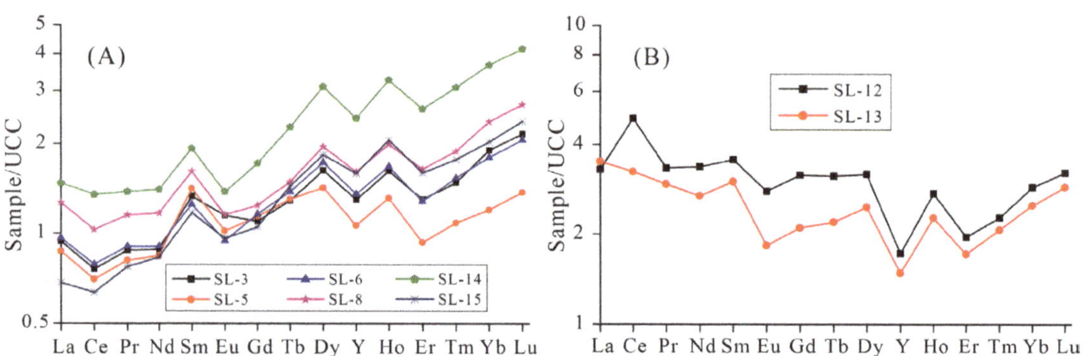

Figure 10. UCC-normalized REY distribution patterns in the studied samples (**A**,**B**). UCC data from Taylor and McLennan [22].

5. Discussion

5.1. The Nature of Detrital Materials

Some previous studies have indicated that the Yunkai Upland located to the east of the Heshan, Yishan, Fusui, and Xian'an Coalfields (Figure 1B) is the dominant sediment source region providing felsic detrital materials into the late Permian coal-bearing basins in Guangxi Province, China [5,6]. However, a few studies have demonstrated that some detrital materials within the late Permian coal-bearing strata in Guangxi Province are probably derived from the basaltic materials from the Kangdian Upland [25], the eroded materials from the felsic igneous rocks at the top of the Kangdian Upland [26,27]. The investigated samples are most likely derived from the eroded felsic detrital materials from the Yunkai Upland based on the following evidence:

(1) The Al_2O_3/TiO_2 ratio has been extensively used to infer the parent rock composition of mudstones [28] and coals [20,29–31], with 3–8, 8–21, and 21–70 indicative of mafic, intermediate, and felsic igneous rocks, respectively [28]. The plot of Al_2O_3 versus TiO_2 shows that the samples studied all fall within the category of felsic rocks (Figure 11A), indicating the input of felsic detrital materials. However, the samples from layers A–C display distinctly low Al_2O_3/TiO_2 (20.6–41.7, 30.9) compared to those from layers D–F (37.7–67.7, 50.6 on average), appearing to indicate that the detrital materials in layers A–C are possibly derived from a mixture of much more felsic constituents and much less mafic materials.

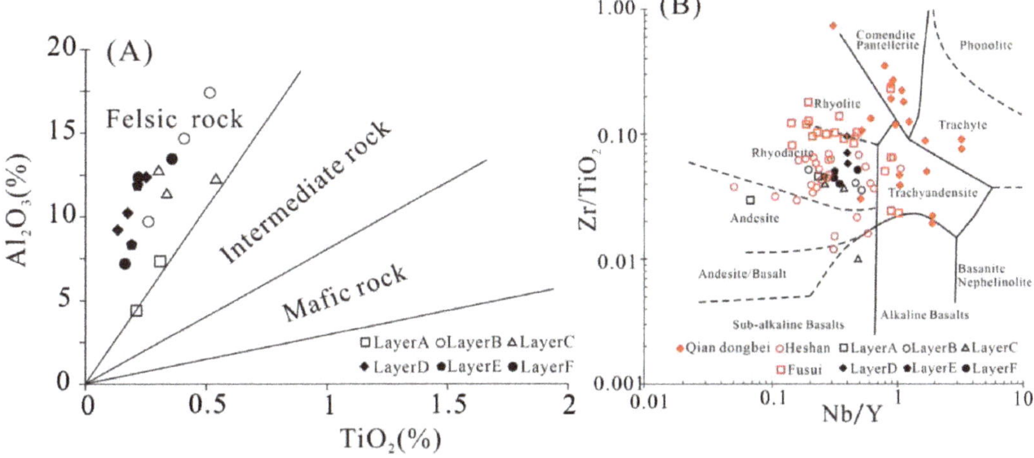

Figure 11. (**A**) the plot of Al$_2$O$_3$ versus TiO$_2$ contents for the investigated samples; (**B**) the plot of Zr/TiO$_2$ versus Nb/Y contents for the investigated samples, the data of Qiandongbei [31], Heshan [3], and Fusui [5] are cited to compare.

(2) The Eu anomaly is also commonly used in monitoring parent rocks. In most cases, the coals with the input of mafic detrital materials display a positive Eu anomaly while the coals with the input of felsic detrital materials are characterized by a negative Eu anomaly [32]. The samples studied display negative Eu anomalies ranging from 0.67 to 0.95 with an average of 0.77, suggesting the input of felsic components. The Eu anomaly and Al$_2$O$_3$/TiO$_2$ ratio in the samples studied appear to exclude the basaltic detrital materials from the Kangdian Upland as the predominant terrigenous sediment region.

(3) In the plot of Nb/Y versus Zr/TiO$_2$ (Figure 11B), the samples studied fall within the field of rhyodacite-dacite, comparable to the late Permian Heshan, Fusui coals in Guangxi Province, China [3,5], indicating that the investigated samples have the same source region to the late Permian coals in Guangxi Province. Moreover, the studied samples differ from the felsic igneous rocks from the Kangdian Upland, which cluster within the field of alkaline rocks [31] (Figure 11B), appearing to exclude the possibility that the felsic constituents are sourced from the felsic rocks at the top of Kangdian Upland.

(4) The UCC-normalized REY distribution patterns in the investigated samples are very similar to that of Heshan coals with terrigenous materials from the Yunkai Upland [3,5] but distinctly differ from that of felsic igneous rocks from the Kangdian Upland [31], further confirming the input of felsic detrital materials from the Yunkai Upland.

(5) The samples studied are primarily composed of quartz, albite, clay minerals, and carbonate minerals. This mineral assemblage would correspond well to the Heshan and Fusui coals with a felsic detritus from the Yunkai Upland [3,5].

5.2. Influence of Seawater Invasion

The coal-bearing layers A–F studied are intercalated within the carbonate successions (Figure 2B), indicating a significant marine invasion during or shortly after coal-bearing deposition. The seawater invasion is also confirmed by geochemical indicators, such as sulfur content, boron content, and Sr/Ba ratio. The previous studies show that the medium- and high-sulfur coals are formed in marine-influenced environments, while the low-sulfur coals are deposited in freshwater-influenced environments [12,33]. The sulfur content in the investigated samples ranges from 2.1 to 8.6% with an average of 4.2% (high-S coal), indicating a significant seawater influence. The boron content in layers A–F range from 45 to 241 µg/g with an average of 100 µg/g, suggesting mildly brackish water-influenced coal-forming environments (B concentrations <50, 50–110, and >110 µg/g indicative of freshwater, mildly brackish water, and brackish water, respectively [34]). The Sr/Ba ratio is considered a useful index of depositional environments with Sr/Ba > 1 and Sr/Ba < 1 indicative of marine-influenced environments and freshwater-influenced environments, respectively [33]. In the present study, the Sr/Ba ratio in layers A–F ranges from 2.6 to 10.6 with an average of 5.9 evidently exceeding 1.0, revealing seawater-dominated-influenced environments, which is consistent with the sedimentary setting of an isolated carbonate platform.

Layers A–C exhibit higher sulfur content (2.8–8.6%, 5.8% on average), Sr/Ba ratio (5.5–10.6, 8.3) than sulfur content (2.1–4.6%, 2.6%), and Sr/Ba ratio (2.4–5.9, 3.5) in layers D–F, indicating a more intensive marine injection during the deposition of layers A–C than layers D–F. The various degrees of marine injection result in different geochemical patterns. For example, layers A–C display higher contents of illite, pyrite, Co, Ni, As, and Se than in layers D–F. This is because Co, Ni, As, and Se are primarily associated with pyrite and the formation of pyrite is intimately associated with the degree of seawater injection [12]. Higher illite abundance in layers A–C is ascribed to the marine-influenced environments because illite is preferentially deposited in brackish-influenced alkali conditions [30,35]. Although the REY content (264 µg/g on average) in layers A–C is higher than that (152 µg/g on average) in layers D–F, seawater is not the primary geological control on REY distribution in layers A–F; otherwise, the pronounced positive Y and Gd anomalies would be expected in layers A–F [36–38], which is in sharp contrast to results shown in Figure 10. Additionally, the Y/Ho molar ratio in the samples ranges from 32.0 to 48.6 with an average of 41.0, which is essentially identical to that in UCC (51.0) but lower than that reported in seawater (90–110; [36]), further indicating a negligible influence of seawater control on the REY content in the investigated samples.

5.3. Mode of Occurrence of Elements

Zirconium and Hf are commonly hosted in zircon [31]. The Zr and Hf in the samples studied show a significant positive correlation (r = 0.95, Figure 12A), indicating that Zr and Hf are hosted in the same mineral carrier and do not evidently fractionate during the formation of coal-bearing strata. The slope of the Zr-Hf regression equation (34.6) is very comparable to that of zircon in granite (38.5, [39]), revealing zircon as the major carrier of Zr and Hf.

The REY contents in the samples from layers C–F display a negative correlation with LTA (r = −0.52, Figure 12B), positive correlation with Zr (r = 0.74, Figure 12C), and slight correlation with Al_2O_3 (r = 0.26) and P (r = 0.14), appearing to indicate that REY is jointly hosted by organic matter and heavy minerals, such as zircon. In most cases, the MREY and HREY are preferentially adsorbed to organic matter, thus leading to the H-type REY distribution patterns in layers C–F (Figure 10A). Additionally, the negative correlation between low-temperature ash yield and (MREY + HREY)/REY (r = −0.52; Figure 12D) indicates that the MREY and HREY comprise more proportions of total REY with increasing organic matter. By contrast, the REY content in the samples from layers A–B is positively correlated with LTA (r = 0.86, Figure 12B), Al_2O_3 (r = 0.99, Figure 12E), and Zr (r = 0.86, Figure 12C), suggesting that REY is predominantly hosted by inorganic matters (e.g., clay minerals, zircon) rather than organic matter. Although the REY in layers A–F shows a relatively significant correlation with Zr, zircon is not the major carrier for REY because the UCC-normalized REY distribution of the samples studied (Figure 10) is not consistent with the REY distribution of zircon (HREY enrichment type, positive Ce anomaly, negative Eu anomaly, and positive Y anomaly [40]). The REY-rich samples SL-12 and SL-13 from layer B display a higher La/Ho ratio compared to other benches (Figure 9), probably indicating precipitation of LREY-enriched mineral phases, such as monazite, which preferentially incorporate the LREE but do not fractionate Y and Ho (a rather constant Y/Ho ratio) as reported by Bau and Dulski [36] and Chesley et al. [41]. Additionally, the L-type REY distribution (Figure 10B) and relatively high P and Th content in the REY-rich samples SL-12 and SL-13 appear to indicate the occurrence of phosphate minerals, such as monazite.

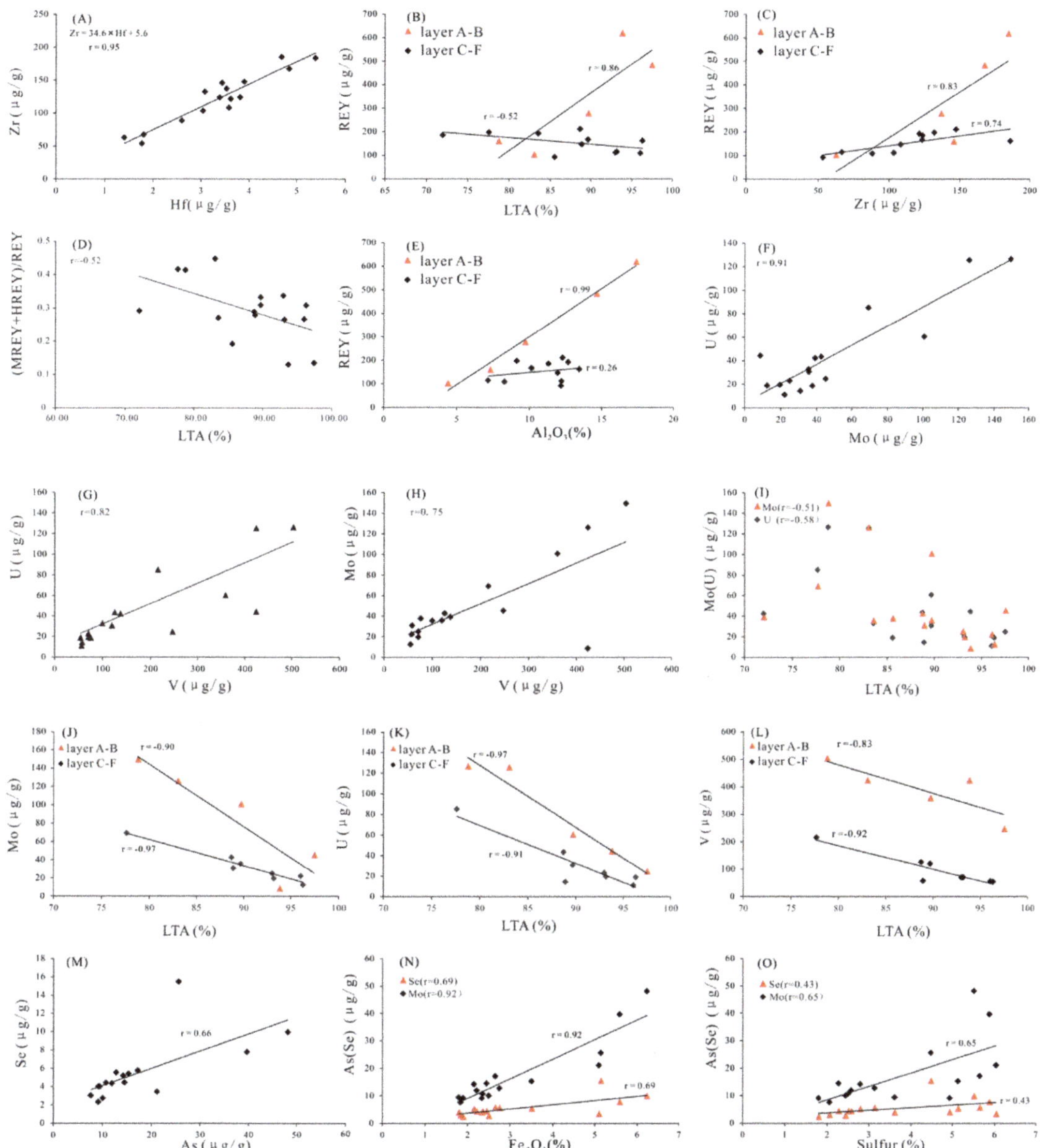

Figure 12. Correlation among selected elements and LTA: (**A**) plot of Zr vs. Hf; (**B**) plot of REY vs. LTA; (**C**) plot of REY vs. Zr; (**D**) plot of (MREY + HREY)/REY vs. LTA; (**E**) plot of REY vs. Al_2O_3; (**F**) plot of U vs. Mo; (**G**) plot of U vs. V; (**H**) plot of Mo vs. V; (**I**) plot of Mo(U) vs. LTA; (**J**) plot of Mo vs. LTA; (**K**) plot of U vs. LTA; (**L**) plot of V vs. LTA; (**M**) plot of Se vs. As; (**N**) plot of As(Se) vs. Fe_2O_3; and (**O**) plot of As(Se) vs. Sulfur.

Mo, U, and V show a similar distribution throughout the borehole profile (Figure 9) and display a positive correlation between Mo and U (r = 0.91, Figure 12F), U and V (r = 0.82, Figure 12G), and Mo and V (r = 0.75, Figure 12H), indicating the same carrier for them or similar geological controls on their enrichment. Mo and U negatively correlate with LTA (Figure 12I) in all samples; in layers A–B and D–F, Mo, U, and V significantly and negatively correlate with LTA (Figure 12J–L), appearing to confirm that Mo, U, and V are predominantly hosted by organic matter, which is consistent with the previous studies indicating that Mo, U, and V in the coals intercalated within carbonate succession commonly show an organic affinity [24,28].

The distribution of As is comparable to that of Se, and they display a relatively significant correlation (r = 0.66, Figure 12M), indicating the same carrier for As and Se. The As and Se positively correlate with Fe_2O_3 (Figure 12N) and sulfur (Figure 12O), suggesting a primary Fe-sulfide affinity for them. The lower correlation coefficient of As–S and Se–S than As–Fe_2O_3 and Se–Fe_2O_3 is ascribed to the additional existence of organic sulfur.

5.4. Elevated Concentrations of Trace Elements

5.4.1. Rare Earth Elements and Yttrium (REY)

The REY content in layer B ranges from 278 to 619 µg/g with an average of 460 µg/g higher than the UCC (168 µg/g; [22]). The highly elevated concentration of REY in layer B is a joint result of the input of alkaline detrital materials and solution/rock interaction based on the following evidence:

(1) As mentioned above, the Yunkai Upland provides felsic detrital materials into the late Permian coal-bearing basin, which do not cause the geochemical anomaly of most lithophile elements in the samples except for those from layer B, appearing to indicate that the terrigenous felsic detrital materials from the Yunkai Upland contribute to normal geochemical background values and other geological factors possibly exert an important influence on REY enrichment in layer B.

(2) The REY-rich samples SL-12 and SL-13 in layer B have higher Nb/Y ratios compared to other bench samples (Figure 11B), appearing to indicate more input of alkaline detrital materials because alkaline rocks compared to sub-alkaline rocks have higher Nb/Y ratio [42]. Moreover, the elevated REY content in layer B is accompanied by elevated concentrations of Zr, Nb, and Ga, which are commonly found in the coals influenced by alkaline volcanic ash [13,24,31], also supporting the input of alkaline pyroclastic materials.

(3) The REY-rich samples SL-12 and SL-13 in layer B are characterized by L-type REY distribution (Figure 10B), which is commonly caused by the input of terrigenous clastic materials or pyroclastic materials [24]. As discussed above, terrigenous felsic detrital input is excluded as a predominant geological factor contributing to the REY enrichment, appearing to indicate that the pyroclastic material is responsible for the L-type REY distribution.

(4) The abundant occurrence of albite and vermicular kaolinite (Figure 6C,F,G), combined with the negative Eu anomaly, appears to indicate the influx of felsic pyroclastic materials because vermiculate kaolinite is commonly derived from the alteration of volcanic ash [43].

(5) The bench sample (sample SL-12) with the highest REY content shows a most notable feature of positive Ce anomaly compared to any of the other benches, revealing that the bench sample is subjected to the oxidation of oxygen-rich solutions, which oxidizes Ce^{3+} to Ce^{4+} that is preferentially adsorbed to clay minerals or deposits as cerianite [44]. In this case, the positive Ce anomaly is most likely caused by preferential adsorption of Ce^{4+} onto clay minerals rather than deposition of cerianite because Ce/La (3.2), Ce/Pr (13.2), and Ce/Nd (3.6) ratio in the bench sample SL-12 is markedly lower than Ce/La (222), Ce/Pr (257), and Ce/Nd (177) in cerianite [45]. The transformation from positive Ce anomaly, no Ce anomaly, to negative Ce anomaly in the upper, middle, and lower portions of layer B, respectively, are due to preferential adsorption of Ce^{4+} onto clay minerals in the upper portion of layer B and, consequently, Ce-poor solutions migrate downward into the lower portion. The lower Y/Ho ratio in the REY-rich benches (17.6) compared to other REY-poor benches (22.9) also confirms the solutions/rocks interaction during which Y and Ho are released, but Ho, relative to Y, is more readily adsorbed by clay minerals [36], causing a relatively low Y/Ho ratio in the REY-rich benches. During the interaction of solution/rock, the HREY are easily released and complexed relative to LREY; the released HREY ions

migrate downward and are ultimately absorbed by clay minerals with the increasing pH due to reaction between HREY-containing solutions with detrital materials, which explains the L-type REY distribution in the upper portion and H-type REY distribution in the lower portion of layer B (Figure 10).

(6) Albite is relatively high in layers D–F, but it is absent in layers A–C, indicating more intensive chemical weathering or solution/rock reaction in layers A–C, which results in the transformation of albite to clay minerals.

5.4.2. Uranium, Mo, V, As, and Se

Uranium, Mo, and V are highly enriched in layer A and layer B and their contents are comparable to the U–Mo–V-rich coals intercalated within carbonate successions [3,5,8,46]. The highly elevated U content is commonly accompanied by Mo and V in the Heshan coals [1–3], Fusui coals [5], Yishan coals [13] (anomaly), and Shanglin coals [6], which were deposited in isolated carbonate platform environments. In the previous study, the highly enriched concentrations of U, Mo, and V are primarily governed by the infiltration of syngenetic or early diagenetic low-temperature hydrothermal fluids rather than the detrital materials from the sediment-source region and marine influence as confirmed by the following evidence:

(1) The felsic detrital materials can be excluded because the felsic rocks are usually depleted in compatible elements, such as V and Cr, which are enriched in the investigated samples (Figure 8).

(2) The previous studies demonstrate that seawater influence may exert an important effect on U and Mo enrichment [47]. In the present study, this is not the case because U (r = −0.67), Mo (r = −0.26), and V (r = −0.86) display a negative correlation with Sr/Ba ratio, which is a useful indicator of seawater invasion [3]. The negative correlation reveals an adverse influence of marine invasion on U, Mo, and V accumulation.

(3) The enhanced concentration of element assemblage of U–Mo–V–As–Se in the samples studied is essentially comparable to that in the coals subjected to the influence of hydrothermal fluids [3], revealing the infiltration of hydrothermal fluids. Moreover, the hydrothermal fluids infiltrating into the investigated samples are regarded as being low-temperature hydrothermal fluids because the high-U and low-U samples show similar negative Eu and Y anomaly, appearing to rule out the injection of high-temperature solutions; otherwise, the positive Eu and Y anomaly can be expected [44]. Furthermore, the low-temperature hydrothermal fluids most likely migrate and permeate into the coal-bearing strata during peat accumulation (syngenetic stage) or shortly after peat accumulation (early diagenetic stage) because the U-rich layers are intercalated among the impermeable clays and limestones, which hamper the infiltration of hydrothermal fluids at the late diagenetic stage.

6. Conclusions

Based on the study of the mineralogy and geochemical characteristics of samples in the studied area, the following conclusions can be drawn:

(1) The minerals in layers A–C consist mainly of quartz, illite, kaolinite, and carbonate phases while those in layers D–F are predominantly composed by quartz, albite, kaolinite, and carbonate phases.

(2) The REY, U, Mo, As, Se, and V are predominantly enriched in layers A and B. The REY is hosted primarily by clay minerals, and U, Mo, and V are primarily associated with organic matter while As and Se show an Fe-sulfide affinity.

(3) The more extensive marine invasion results in the higher contents of pyrite, sulfur, As, Se, Co, and Ni in layers A–C than layers D–F; the input of alkaline pyroclastic materials and the interaction of O_2-rich solutions and detrital materials jointly govern the REY enrichment and distribution pattern; the influx of low-temperature hydrothermal fluids at the syngenetic or early diagenetic stage is the predominant source of U, Mo, As, Se, and V in layers A–B. The elements U, Mo, V, Se, and REY are highly promising for recovery.

Author Contributions: B.L. (Bo Li), B.L. (Baoqing Li), X.Z., F.Z., X.Q., Y.S. and J.L. collected the samples; X.Q. and N.M. conducted the experiments; B.L. (Bo Li) wrote the original draft; B.L. (Baoqing Li) revised and edited the manuscript. All authors have read and agreed to the published version of the manuscript.

Funding: This research was funded by the National Key Research and Development Program of China (No. 2021YFC2902005), National Science Foundation of China (No. 41972182), the National Science Foundation of Guangxi Province (No. 2018JJA150165), and Science Program of China National Administration of Coal Geology (no. ZMKJ-2021-ZX03).

Acknowledgments: The authors would like to give their sincere thanks to Guangxi Bureau of Coal Geology for assistance during sampling and Institute of Environmental Assessment and Water Research, CSIC, Spain, for assistance during the sample analysis.

Conflicts of Interest: The authors declare no conflict of interest.

References

1. Shao, L.Y.; Jones, T.; Gayer, R.; Dai, S.F.; Li, S.S.; Jiang, Y.F.; Zhang, P.F. Petrology and geochemistry of the high-sulphur coals from the Upper Permian carbonate coal measures in the Heshan Coalfield, southern China. *Int. J. Coal Geol.* **2003**, *55*, 1–26. [CrossRef]
2. Zeng, R.S.; Zhuang, X.G.; Koukouzas, N.; Xu, W.D. Characterization of trace elements in sulphur-rich Late Permian coals in the Heshan coal field, Guangxi, South China. *Int. J. Coal Geol.* **2005**, *61*, 87–95. [CrossRef]
3. Dai, S.F.; Zhang, W.G.; Seredin, V.V.; Ward, C.R.; Hower, J.C.; Song, W.J.; Wang, X.B.; Li, X.; Zhao, L.X.; Kang, H.; et al. Factors controlling geochemical and mineralogical compositions of coals preserved within marine carbonate successions: A case study from the Heshan Coalfield, southern China. *Int. J. Coal Geol.* **2013**, *109*, 77–100. [CrossRef]
4. Zhao, Y.Y.; Liang, H.Z.; Zeng, F.G.; Tang, Y.G.; Liang, L.T.; Takahashi, F. Origins and occurrences of Ti-nanominerals in a superhigh-organic-sulfur coal. *Fuel* **2020**, *259*, 116302. [CrossRef]
5. Dai, S.F.; Zhang, W.G.; Ward, C.R.; Seredin, V.V.; Hower, J.C.; Li, X.; Song, W.J.; Wang, X.B.; Kang, H.; Zheng, L.C.; et al. Mineralogical and geochemical anomalies of late Permian coals from the Fusui Coalfield, Guangxi Province, southern China: Influences of terrigenous materials and hydrothermal fluids. *Int. J. Coal Geol.* **2013**, *105*, 60–84. [CrossRef]
6. Zhang, F.Q.; Li, B.Q.; Zhuang, X.G.; Xavier, Q.; Natalia, M.; Shangguan, Y.F.; Zhou, J.M.; Liao, J.L. Geological Controls on Enrichment of Rare Earth Elements and Yttrium (REY) in Late Permian Coals and Non-Coal Rocks in the Xian'an Coalfield, Guangxi Province. *Minerals* **2021**, *11*, 301. [CrossRef]
7. Dai, S.F.; Ji, D.P.; Ward, C.R.; French, D.; Hower, J.C.; Yan, X.Y.; Wei, Q. Mississippian anthracites in Guangxi Province, southern China: Petrological, mineralogical, and rare earth element evidence for high-temperature solutions. *Int. J. Coal Geol.* **2018**, *197*, 84–114. [CrossRef]
8. Dai, S.F.; Seredin, V.V.; Ward, C.R.; Hower, J.C.; Xing, Y.W.; Zhang, W.G.; Song, W.J.; Wang, P.P. Enrichment of U-Se-Mo-Re-V in coals preserved within marine carbonate successions: Geochemical and mineralogical data from the Late Permian Guiding Coalfield, Guizhou, China. *Min. Depos.* **2015**, *50*, 159–186. [CrossRef]
9. Dai, S.F.; Ren, D.Y.; Zhou, Y.P.; Chou, C.L.; Wang, X.B.; Zhao, L.; Zhu, X.W. Mineralogy and geochemistry of a superhigh-organic-sulfur coal, Yanshan Coalfield, Yunnan, China: Evidence for a volcanic ash component and influence by submarine exhalation. *Chem. Geol.* **2008**, *255*, 182–194. [CrossRef]
10. Li, W.W.; Tang, Y.G. Sulfur isotopic composition of superhigh-organic-sulfur coals from the Chenxi coalfield, southern China. *Int. J. Coal Geol.* **2014**, *127*, 3–13. [CrossRef]
11. Chou, C.L. Geologic factors affecting the abundance, distribution, and speciation of sulfur in coals. In *Geology of Fossil Fuels—Coal*; CRC Press: Boca Raton, FL, USA, 1997.
12. Chou, C.L. Sulfur in coals: A review of geochemistry and origins. *Int. J. Coal Geol.* **2012**, *100*, 1–13. [CrossRef]
13. Dai, S.F.; Xie, P.P.; Ward, C.R.; Yan, X.Y.; Guo, W.M.; French, D.; Graham, I.T. Anomalies of rare metals in Lopingian super-high-organic-sulfur coals from the Yishan Coalfield, Guangxi, China. *Ore Geol. Rev.* **2017**, *88*, 235–250. [CrossRef]
14. Dai, S.F.; Ren, D.Y.; Chou, C.L.; Finkelman, R.B.; Seredin, V.V.; Zhou, Y.P. Geochemistry of trace elements in Chinese coals: A review of abundances, genetic types, impacts on human health, and industrial utilization. *Int. J. Coal Geol.* **2012**, *94*, 3–21. [CrossRef]
15. Liao, J.L.; Zhang, F.Q.; Wei, S.M.; Liang, X.D. Lithium and gallium abundance and enrichment factors in typical Late Permian coal-accumulating basin in Guangxi. *Coal Geol. Explor.* **2020**, *48*, 1.
16. Yang, H.Y. *Devonian-Middle Triassic Tectono-Palaeogeographic Pattern and Its Evolution in Hunan-Guangxi Area*; China University of Petroleum (East China): Beijingm, China, 2010.
17. *Standard D3175-11*; Standard Test Method for Volatile Matter in the Analysis Sample of Coal and Coke. ASTM International: West Conshohocken, PA, USA, 2011.
18. *Standard D3174-11*; Annual Book of ASTM Standards. Test Method for Ash in the Analysis Sample of Coal and Coke. ASTM International: West Conshohocken, PA, USA, 2011.

19. *Standard D3173-11*; Standard Test Method for Moisture in the Analysis Sample of Coal and Coke. ASTM International: West Conshohocken, PA, USA, 2011.

20. Li, B.Q.; Zhuang, X.G.; Querol, X.; Moreno, N.; Cordoba, P.; Li, J.; Zhou, J.B.; Ma, X.P.; Liu, S.B.; Shangguan, Y.F. The mode of occurrence and origin of minerals in the Early Permian high-rank coals of the Jimunai depression, Xinjiang Uygur Autonomous Region, NW China. *Int. J. Coal Geol.* **2019**, *205*, 58–74. [CrossRef]

21. Querol, X.; Whateley, M.K.G.; FernandezTuriel, J.L.; Tuncali, E. Geological controls on the mineralogy and geochemistry of the Beypazari lignite, central Anatolia, Turkey. *Int. J. Coal Geol.* **1997**, *33*, 255–271. [CrossRef]

22. Taylor, S.R.; Mclennan, S.M. The continental crust: Its composition and evolution. *J. Geol.* **1985**, *94*, 57–72.

23. Ketris, M.P.; Yudovich, Y.E. Estimations of Clarkes for Carbonaceous biolithes: World averages for trace element contents in black shales and coals. *Int. J. Coal Geol.* **2009**, *78*, 135–148. [CrossRef]

24. Seredin, V.V.; Dai, S.F. Coal deposits as potential alternative sources for lanthanides and yttrium. *Int. J. Coal Geol.* **2012**, *94*, 67–93. [CrossRef]

25. Huang, H.; Du, Y.S.; Yang, J.H.; Zhou, L.; Hu, L.S.; Huang, H.W.; Huang, Z.Q.J.L. Origin of Permian basalts and clastic rocks in Napo, Southwest China: Implications for the erosion and eruption of the Emeishan large igneous province. *Lithos* **2014**, *208–209*, 324–338. [CrossRef]

26. Yang, J.; Cawood, P.A.; Du, Y.J.E.; Letters, P.S. Voluminous silicic eruptions during late Permian Emeishan igneous province and link to climate cooling. *Earth Planet. Sci. Lett.* **2015**, *432*, 166–175. [CrossRef]

27. Deng, J.; Wang, Q.; Yang, S.; Liu, X.; Zhang, Q.; Yang, L.; Yang, Y. Genetic relationship between the Emeishan plume and the bauxite deposits in Western Guangxi, China: Constraints from U–Pb and Lu–Hf isotopes of the detrital zircons in bauxite ores—ScienceDirect. *J. Asian Earth Sci.* **2010**, *37*, 412–424. [CrossRef]

28. Hayashi, T.; Hinoda, Y.; Takahashi, T.; Adachi, M.; Miura, S.; Izumi, T.; Kojima, H.; Yano, S.; Imai, K.J.I.M. Idiopathic CD4+ T-lymphocytopenia with Bowen's disease. *Intern. Med.* **1997**, *36*, 822–824. [CrossRef] [PubMed]

29. Hower, J.C.; Dai, S.F. Petrology and chemistry of sized Pennsylvania anthracite, with emphasis on the distribution of rare earth elements. *Fuel* **2016**, *185*, 305–315. [CrossRef]

30. Li, B.Q.; Zhuang, X.G.; Li, J.; Querol, X.; Font, O.; Moreno, N. Geological controls on mineralogy and geochemistry of the Late Permian coals in the Liulong Mine of the Liuzhi Coalfield, Guizhou Province, Southwest China. *Int. J. Coal Geol.* **2016**, *154*, 1–15. [CrossRef]

31. Li, B.Q.; Zhuang, X.G.; Querol, X.; Moreno, N.; Zhang, F. Geological controls on the distribution of REY-Zr (Hf)-Nb (Ta) enrichment horizons in late Permian coals from the Qiandongbei Coalfield, Guizhou Province, SW China. *Int. J. Coal Geol.* **2020**, *231*, 103604. [CrossRef]

32. Dai, S.F.; Chekryzhov, I.Y.; Seredin, V.V.; Nechaev, V.P.; Graham, I.T.; Hower, J.C.; Ward, C.R.; Ren, D.Y.; Wang, X.B. Metalliferous coal deposits in East Asia (Primorye of Russia and South China): A review of geodynamic controls and styles of mineralization. *Gondwana Res.* **2016**, *29*, 60–82. [CrossRef]

33. Dai, S.F.; Bechtel, A.; Eble, C.F.; Flores, R.M.; French, D.; Graham, I.T.; Hood, M.M.; Hower, J.C.; Korasidis, V.A.; Moore, T.A.; et al. Recognition of peat depositional environments in coal: A review. *Int. J. Coal Geol.* **2020**, *219*, 103383.

34. Golab, A.N.; Carr, P.F. Changes in geochemistry and mineralogy of thermally altered coal, Upper Hunter Valley, Australia. *Int. J. Coal Geol.* **2004**, *57*, 197–210. [CrossRef]

35. Rimmer, S.M.; Davis, A. Geologic Controls on the Inorganic Composition of Lower Kittanning Coal. In *Geology of Fossil Fuels—Coal*; ACS Publication: Washington, DC, USA, 1986; pp. 41–52.

36. Bau, M.; Dulski, P. Petrology, Comparative study of yttrium and rare-earth element behaviours in fluorine-rich hydrothermal fluids. *Contrib. Mineral. Petrol.* **1995**, *119*, 213–223. [CrossRef]

37. Bau, M.; Koschinsky, A.; Dulski, P.; Hein, J.R. Comparison of the partitioning behaviours of yttrium, rare earth elements, and titanium between hydrogenetic marine ferromanganese crusts and seawater. *Geochim. Cosmochim. Acta* **1996**, *60*, 1709–1725. [CrossRef]

38. Byrne, R.H.; Kim, K.-H. Rare earth element scavenging in seawater. *Geochim. Cosmochim. Acta* **1990**, *54*, 2645–2656. [CrossRef]

39. Xu, C.; Kynicky, J.I.; Smith, M.P.; Kopriva, A.; Brtnicky, M.; Urubek, T.; Yang, Y.; Zhao, Z.; He, C.; Song, W.J.N.C. Origin of heavy rare earth mineralization in South China. *Nat. Commun.* **2017**, *8*, 14598. [CrossRef] [PubMed]

40. Braun, J.J.; Riotte, J.; Battacharya, S.; Violette, A.; Oliva, P.; Prunier, J.; Subramanian, S. REY-Th-U Dynamics in the Critical Zone: Combined Influence of Reactive Bedrock Accessory Minerals, Authigenic Phases, and Hydrological Sorting (Mule Hole Watershed, South India). *Geochem. Geophys. Geosyst.* **2018**, *19*, 1611–1635. [CrossRef]

41. Chesley, S.; Lumpkin, M.; Schatzki, A.; Galpern, W.R.; Greenblatt, D.J.; Shader, R.I.; Miller, L.G. Prenatal exposure to benzodiazepine—I. Prenatal exposure to lorazepam in mice alters open-field activity and GABAA receptor function. *Neuropharmacology* **1991**, *30*, 53–58. [CrossRef]

42. Winchester, J.A.; Floyd, P.A. Geochemical discrimination of different magma series and their differentiation products using immobile elements. *Chem. Geol.* **1977**, *20*, 325–343. [CrossRef]

43. Dai, S.F.; Ward, C.R.; Graham, I.T.; French, D.; Hower, J.C.; Zhao, L.; Wang, X.B. Altered volcanic ashes in coal and coal-bearing sequences: A review of their nature and significance. *Earth-Sci. Rev.* **2017**, *175*, 44–74. [CrossRef]

44. Dai, S.F.; Graham, I.T.; Ward, C.R. A review of anomalous rare earth elements and yttrium in coal. *Int. J. Coal Geol.* **2016**, *159*, 82–95. [CrossRef]

45. Ichimura, K.; Sanematsu, K.; Kon, Y.; Takagi, T.; Murakami, T.J.A.M. REE redistributions during granite weathering: Implications for Ce anomaly as a proxy for paleoredox states. *Am. Mineral.* **2020**, *105*, 848–859. [CrossRef]
46. Dai, S.; Finkelman, R.B. Coal geology in China: An overview. *Int. Geol. Rev.* **2018**, *60*, 531–534. [CrossRef]
47. Li, B.Q.; Zhuang, X.G.; Li, J.; Querol, X.; Font, O.; Moreno, N. Enrichment and distribution of elements in the Late Permian coals from the Zhina Coalfield, Guizhou Province, Southwest China. *Int. J. Coal Geol.* **2017**, *171*, 111–129. [CrossRef]

Article

Mineralogy and Geochemistry of the Lower Cretaceous Coals in the Junde Mine, Hegang Coalfield, Northeastern China

Yingchun Wei [1,2], Wenbo He [2], Guohong Qin [3,*], Anmin Wang [2] and Daiyong Cao [1]

[1] State Key Laboratory of Coal Resources and Safe Mining, China University of Mining and Technology, Beijing 100083, China; wyc@cumtb.edu.cn (Y.W.); cdy@cumtb.edu.cn (D.C.)
[2] College of Geoscience and Surveying Engineering, China University of Mining and Technology, Beijing 100083, China; hewenbo2020@163.com (W.H.); wamcumtb@163.com (A.W.)
[3] School of Geographic Sciences/Hebei Key Laboratory of Environmental Change and Ecological Construction, Hebei Normal University, Shijiazhuang 050024, China
* Correspondence: qinguohong1@163.com

Abstract: Hegang coalfield is one of the areas with abundant coal resources in Heilongjiang Province. Characteristics of minerals and geochemistry of No. 26 coal (lower Cretaceous coals) from Junde mine, Hegang coalfield, Heilongjiang province, China, were reported. The results showed that No. 26 coal of Junde mine is slightly enriched in Cs, Pb, and Zr compared with world coals. The minerals in No. 26 coal of Junde mine primarily include clay minerals and quartz, followed by calcite, siderite, pyrite, monazite, and zircon. The diagrams of Al_2O_3–TiO_2, Zr/Sc–Th/Sc, Al_2O_3/TiO_2–Sr/Y, and Al_2O_3/TiO_2–La/Yb indicate that the enriched elements in No. 26 coal were mainly sourced from the Late Paleozoic meta-igneous rocks in Jiamusi block. The volcanic ash contribution to No. 26 coal seems very low. Sulfate sulfur indicating oxidation/evaporation gradually decreases during No. 26 coal formation.

Keywords: trace elements; minerals; source; volcanic; sedimentary environment

Citation: Wei, Y.; He, W.; Qin, G.; Wang, A.; Cao, D. Mineralogy and Geochemistry of the Lower Cretaceous Coals in the Junde Mine, Hegang Coalfield, Northeastern China. *Energies* **2022**, *15*, 5078. https://doi.org/10.3390/en15145078

Academic Editor: Dameng Liu

Received: 21 June 2022
Accepted: 11 July 2022
Published: 12 July 2022

Publisher's Note: MDPI stays neutral with regard to jurisdictional claims in published maps and institutional affiliations.

1. Introduction

Coal is considered an important source of critical metals [1,2]. The critical metals in coal have become a research hotspot in recent years due to their potential economic significance [2,3] and the geological implications for coal basins [4,5]. Coals with the significant economic value of critical element production are usually called 'metalliferous coal' [2,6] or 'coal-hosted rare-metal deposit' [7]. More recently, further studies have paid close attention to the occurrence and recovery methods from these coals [8–19].

Heilongjiang Province, Northeastern (EN) China, has a vast territory and rich resources. In terms of coal resources, the Hegang coalfield is rich [20,21]. Although there has been much literature focusing on coal geochemical and mineralogical characteristics in China, these works are mainly concentrated in the coals in southwestern and northern China [13,22–29]. In Heilongjiang province, a few studies [30,31] have been carried out about coal-bearing sequences in the Hegang coalfield. Previous investigations have shown that the Hegang coalfield was affected by three periods of tectonic stress and accompanied by multiple periods of volcanic activity, which had an impact on coal quality in the coal basin [32]. There are still two problems to be solved about coal geochemistry in the Hegang coalfield: (1) did the volcanic activity influence the element enrichment in the coals? (2) The occurrence modes and the provenance of elements in the Palaeogene coal-bearing seams have been deeply studied recently in Jilin province [33–36], neighboring the Heilongjiang province. Is the provenance of trace elements in Hegang coals the same as in the coals from Jilin Province [33,35]? Further investigation of these issues has important theoretical and economic value for improving the comprehensive utilization efficiency of coal measures mineral resources.

To address this issue, we study the minerals and elements characteristics of No. 26 coal in the Hegang coalfield. We also discussed the geological factors influencing element enrichment.

2. Geological Setting

Junde mine is situated in Hegang coalfield, Heilongjiang Province (Figure 1). Geologically, the Hegang coalfield is located on the Jiamusi block inside the Central Asian Orogenic Belt between the Chinese plate and the Siberia plate, and is generally characterized as a semi-covered monoclinic structure dipping eastward. The Hegang coalfield is an important part of the Mesozoic-Cenozoic basin group in the east of Heilongjiang. There are mainly fault structures in the Hegang coalfield, and there are relatively gentle folds in the local area, accompanied by multiple volcanic structures. Proterozoic, Mesozoic and Cenozoic strata developed in the study area [37,38].

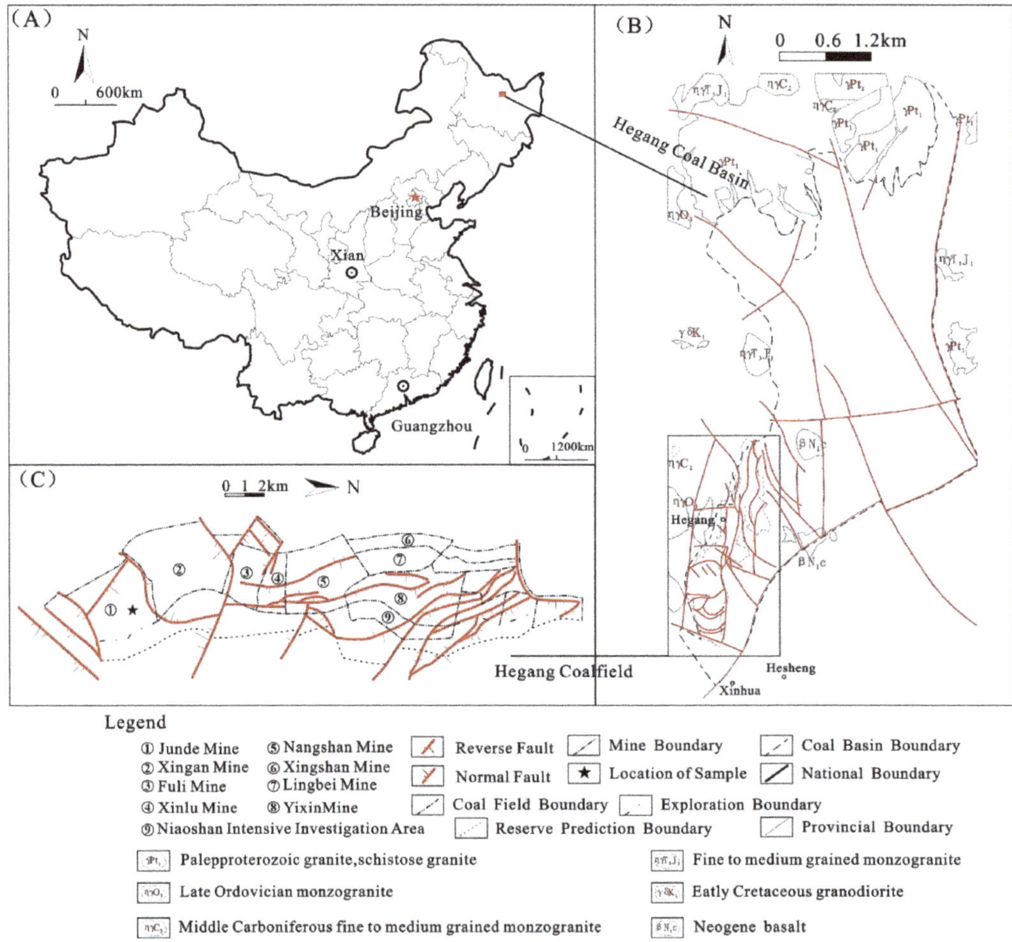

Figure 1. (**A,B**) Site of the Hegang coalfield. (**C**) Location of the Junde mine.

The major coal-bearing formation in the Hegang coalfield is the Lower Cretaceous Chengzihe Formation. The Lower Cretaceous Chengzihe Formation consists of gray-white conglomerate, coarse and fine sandstone, gray-yellow medium sandstone, dark gray siltstone, mudstone, coal seams, and tuff. In Hegang coalfield, 36 coal seams are minable or partially minable, which are divided into four coal bearing groups from top to bottom. The first coal seam group includes coal seams No. 1 and No. 2, which are unstable. The second coal seam group includes coal seams No. 3~22. The third coal seam group, including coal seams No. 23~26, is mainly developed in the south of the coalfield, and its thickness gradually thins or even pinches out to the north. The fourth coal seam group includes coal seams No. 27~32.

No. 26 coal, as the main mining seam in Junde mine from Hegang coalfield, is the main research object of this paper. The thickness of No. 26 coal is about 2.9 m, deposited in continental facies [37,38].

3. Sampling and Analytical Techniques

Ten samples were obtained in No. 26 coal in Junde mine, Hegang Coalfield, which include one roof sample, eight coal samples, and one floor sample (Figure 2). From top down, eight coal samples are identified as JD-26-01 to JD-26-08 (Figure 2B). The roof and floor are identified as JD-26-R and JD-26-F, respectively (Figure 2B). All samples are fresh, without pollution and oxidation.

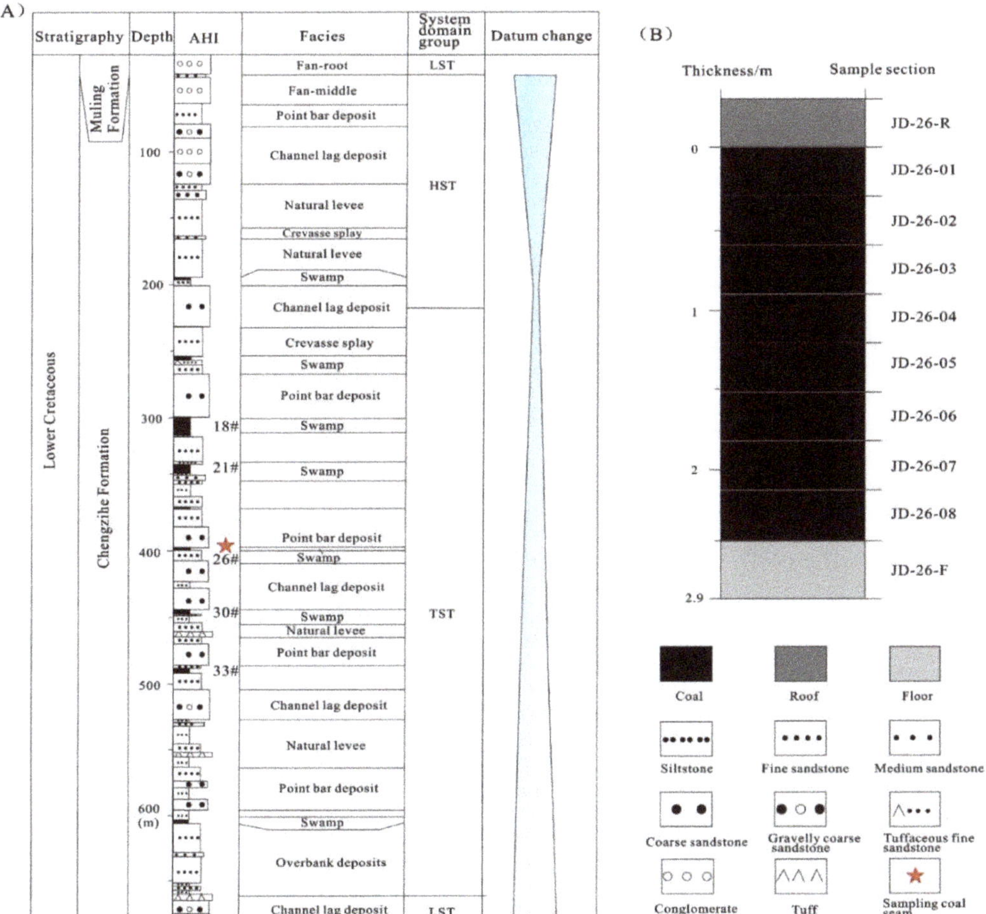

Figure 2. (**A**) The strata histogram (modified after [38]) and (**B**) sampling profile from Junde mine, Hegang coalfield.

According to ASTM D3173-11, D3174-11 and D3175-11(2011) [39–41], the coal samples collected from Junde mine were tested for the proximate analysis. The macerals were obtained based on the ICCP System 1994 (ICCP, 1998, 2001) [42,43] and Pickel, Kus [44]. The total sulfur content was obtained by ZCL-3 automatic sulfur analyzer, and the contents of various sulfur forms in the samples were obtained according to ASTM D3177-02(2011) and ASTM D2492-02(2012) [45,46]. These analyses were made at the Société Générale de Surveillance S.A-China Stand Science and Technology Group Corporation (SGS-CSTC) Standards Technology Service Corporation.

Based on SY/T 5163-2010, the minerals were identified by X-ray diffractometer. The morphology of minerals was observed by the scanning electron microscopy (SEM) with an energy dispersive X-ray spectrometer (EDS). The working distance was 8.4–11.4 mm, and beam voltage is 15.0 kV during the SEM operation. These experiments were conducted at the Beijing Center for Physical and Chemical Analysis (BCPCA).

The cleaned samples used for the geochemical analysis were crushed and ground to less than 200-mesh size using a tungsten carbide ball mill. The Major Oxides contents were obtained using X-ray fluorescence spectrometry (XRF; PANalytical Axios). The trace elements were obtained by the inductively coupled plasma mass spectrometry (ICP-MS, ELEMENT XR). Before ICP-MS analysis, samples were digested by using an UltraClave Microwave High Pressure Reactor (Milestone, Milan, Italy). Multi-element standards (Inorganic Ventures: CCS-1, CCS-4, CCS-5, and CCS-6; GBW07103, GBW07104) were used for trace element concentrations calibration. More method details can be seen in [47]. XRF analyses and the ICP-MS analyses were performed at the China National Nuclear Corporation (CNNC) Beijing Research Institute of Uranium Geology.

In our study, concentration coefficients (CC) proposed by [48] is adopted. CC = ratio of the element contents in the coals/clays to average contents in worldwide coals/clays. On this basis, the concentrations of elements can be classified into six categories, i.e., unusually enriched (CC > 100), significantly enriched (10 < CC < 100), enriched (5 < CC < 10), slightly enriched (2 < CC < 5), normal (0.5 < CC < 2), and depleted (CC < 0.5).

4. Results

4.1. Coal Characteristics and Coal Petrology

As shown in Table 1, the ash yield and total sulfur of Junde coals is 8.6–17.6% (13.0% on average) and 0.54–0.72%, (0.62% on average). The contents above show that No. 26 coals are characterized by low ash and low sulfur based on GB15224.1-2010 and GB15224.2-2010 [49,50]. In addition, sulfur in coal occurs mainly in organic and pyritic sulfur. Sulfate sulfur increases with depth.

Table 1. Proximate analysis and sulfur contents (%) of coals from Junde mine.

Sample	M_{ad}	A_d	V_{daf}	Total Sulfur	Pyritic Sulfur	Sulfate Sulfur	Organic Sulfur
JD-26-01	3.1	12.6	36.2	0.54	0.27	0.04	0.23
JD-26-02	2.4	13.4	37.2	0.57	0.02	0.05	0.50
JD-26-03	2.4	17.5	38.8	0.57	0.02	0.07	0.48
JD-26-04	2.2	12.1	38.6	0.72	0.43	0.08	0.20
JD-26-05	2.3	8.6	37.3	0.58	0.31	0.07	0.20
JD-26-06	2.3	9.7	37.9	0.66	0.33	0.09	0.24
JD-26-07	1.9	12.4	38.3	0.62	0.35	0.13	0.14
JD-26-08	2.2	17.6	41.1	0.70	0.44	0.14	0.11

M_{ad}—air-dry basis moisture; A_d—dry basis ash yield; V_{daf}—dry and ash-free basis volatile matter.

The vitrinite (84.8 vol.% on average) is the most abundant macerals in No. 26 coal (Table 2). It is composed mainly of collotelinite (Figure 3A–C,E,I) and collodetrinite (Figure 3A,D,I), and to a lesser extent, by telinite (Figure 3H), with a few corpogelinite and virtrodetrinite. The collodetrinite is usually in matrix form, embedded in quartz (Figure 3B), clay minerals (Figure 3D,I), calcite (Figure 3E), or fracture-filling pyrite (Figure 3G).

Table 2. Coal petrology characteristics of No. 26 coals (Vol.%).

Sample	Vitrinite						Inertinite					Liptinite						Mineral			
	T	CT	CD	VD	CG	T-V	F	SF	Mic	ID	T-I	Sp	Cut	Sub	Res	Bit	T-L	Clays	Sulfide Minerals	Carbonate Minerals	Silica Minerals	Total Minerals
JD-26-01	11.1	38.7	26.6	0.5	1.0	77.9	0.5	14.6	-	2.0	17.1	0.5	0.5	-	-	0.5	1.5	0.5	1.0	1.5	0.5	3.5
JD-26-02	19.2	29.8	31.8	0.5	0.5	81.8	0.5	7.6	-	1.0	9.1	1.5	-	-	0.5	-	2.0	2.5	-	3.5	1.0	7.1
JD-26-03	10.5	37.1	38.1	-	1.9	87.6	1.9	3.8	-	1.9	7.6	1.0	0.5	-	-	-	1.4	2.4	-	-	1.0	3.3
JD-26-04	8.0	40.3	34.3	1.0	2.0	85.6	0.5	4.5	0.5	1.0	6.5	1.5	-	-	0.5	0.5	2.5	2.5	0.5	1.0	1.5	5.5
JD-26-05	11.0	49.8	27.8	-	3.8	92.3	-	3.4	-	0.5	3.8	1.0	0.5	-	-	-	1.4	-	0.5	1.9	-	2.4
JD-26-06	15.1	40.6	31.8	-	3.7	91.2	1.0	4.7	0.5	-	6.3	0.5	-	0.5	-	-	1.0	0.5	-	1.0	-	1.6
JD-26-07	9.5	39.5	35.0	-	2.0	86.0	3.0	5.0	-	0.5	8.5	1.0	0.5	-	0.5	-	2.0	1.5	-	1.0	1.0	3.5
JD-26-08	6.6	26.5	39.8	-	3.1	76.0	0.5	4.1	-	2.0	6.6	4.6	-	-	-	1.0	5.6	4.6	-	5.6	1.5	11.7

Notes: T—telinite; CT—collotelinite; CD—collodetrinite; CG—corpogelinite; VD—Virtrodetrinite; T-V—total vitrinites; F—fusinite; SF—semifusinite; Mic—micrinite; ID—inertodetrinite; T-I—total inertinites; Sp—sporinite; Cut—cutinite; Res—resinite; Sub—suberinite; Bit—bituminite; T-L—total liptinites; -, no test or no result.

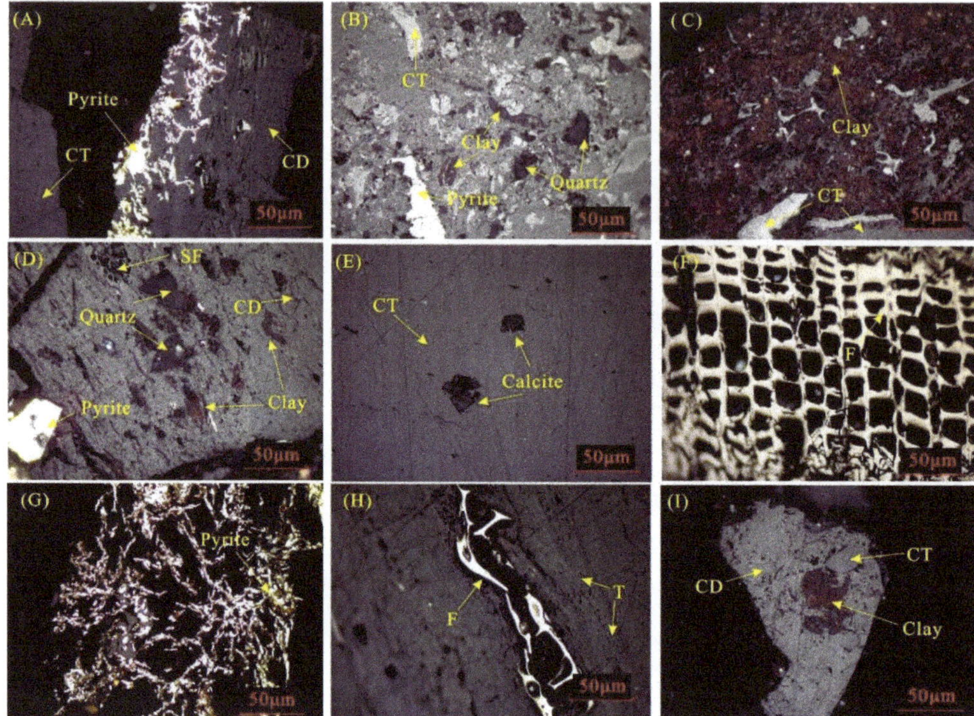

Figure 3. Coal petrology characteristics in No. 26 coal. (**A**) Collotelinite, collodetrinite, and pyrite in sample JD-26-01; (**B**) Collotelinite, clay, quartz, and pyrite in sample JD-26-01; (**C**) Collotelinite and clays in sample JD-26-04; (**D**) Collodetrinite, semifusinite, quartz, pyrite and clay minerals in sample JD-26-04; (**E**) Collotelinite and calcite in sample JD-26-05; (**F**) Fusinite in sample JD-26-07; (**G**) Pyrite in in sample JD-26-07; (**H**) Telinite and Fusinite in sample JD-26-07; (**I**) Collodetrinite, collotelinite and clays in sample JD-26-08. T—telinite; CT—collotelinite; CD—collodetrinite; F—fusinite; SF—semifusinite.

The inertinite (8.2 vol.% on average) is the second most abundant macerals in No. 26 coal in the Junde mine (Table 2). It primarily consists of semifusinite (Figure 3D), inertodetrinite, and fusinite (Figure 3F,H), with small amounts of micrinite. In the majority of cases, the cell walls of fusinite and semifusinite are swollen and not intact, indicating the existence of degradation [51,52] (Figure 3D,F,H).

The liptinite (2.2 vol.% on average) is mainly represented by sporinite (1.4 vol.% on average), with trace amounts of cutinite (0.3 vol.% on average), bituminite (0.3 vol.% on average), resinite (0.2 vol.% on average), and suberinite (0.1 vol.% on average).

4.2. Geochemical Features

4.2.1. Major Oxides (MEO)

Table 3 shows the MEO percentages of studied coals and the average value of Chinese coals. The MEO in studied coals are primarily composed of SiO_2 (average 7.63%), Al_2O_3 (average 2.95%) and Fe_2O_3 (average 1.62%). By comparison with the average values of Chinese coals [22], the percentages of MnO and K_2O are a bit higher, while the content of other MEO is near or below (Figure 4). Value of SiO_2/Al_2O_3 of No. 26 coals (average 2.65) exceeds that of Chinese coals [22] and the theoretical figure of SiO_2/Al_2O_3 of kaolinite, probably because a high content of quartz existed in the studied coals (Figure 3B,D).

Table 3. The concentration of MEO in No. 26 coals, on a whole coal basis (%).

Sample	LOI	SiO$_2$	TiO$_2$	Al$_2$O$_3$	Fe$_2$O$_3$	MgO	CaO	MnO	Na$_2$O	K$_2$O	SiO$_2$/Al$_2$O$_3$	Al$_2$O$_3$/TiO$_2$
JD-26-01	87.41	7.35	0.08	2.99	0.99	0.22	0.56	0.03	0.03	0.33	2.46	36.93
JD-26-02	86.65	7.75	0.08	3.01	1.41	0.22	0.49	0.04	0.04	0.3	2.57	39.15
JD-26-03	82.47	10.94	0.11	4.76	1.04	0.15	0.09	0.01	0.05	0.38	2.3	42.61
JD-26-04	87.93	7.64	0.06	2.63	1.32	0.12	0.07	0.02	0.03	0.16	2.9	45.15
JD-26-05	91.37	4.8	0.05	1.64	1.68	0.16	0.11	0.04	0.02	0.11	2.93	29.92
JD-26-06	90.35	5.44	0.07	1.85	1.76	0.16	0.14	0.07	0.03	0.14	2.93	28.48
JD-26-07	87.64	7.52	0.1	2.88	1.38	0.16	0.07	0.05	0.04	0.16	2.61	28.1
JD-26-08	82.38	9.56	0.15	3.83	3.35	0.25	0.06	0.14	0.04	0.25	2.5	25.96
Average		7.63	0.09	2.95	1.62	0.18	0.20	0.05	0.04	0.23	2.65	34.54
China [a]		8.47	0.33	5.98	4.85	0.22	1.23	0.02	0.16	0.19	1.42	18.12

Notes: LOI, loss on ignition; [a], concentration of MEO in Chinese coals [22].

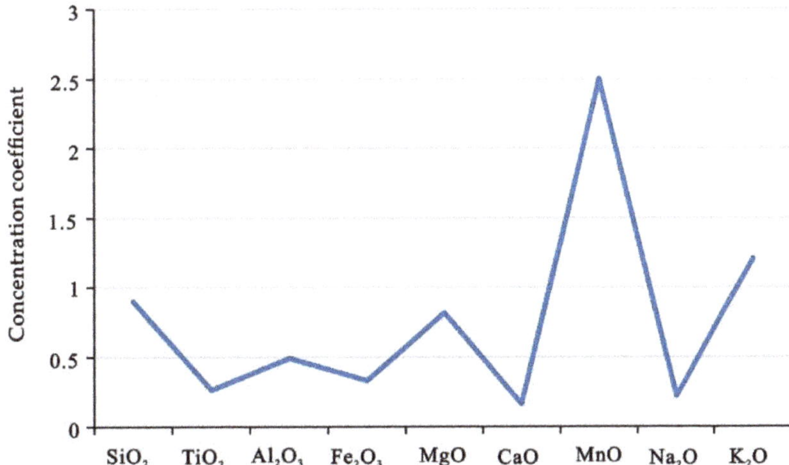

Figure 4. Concentration coefficient (CC) of MEO in No. 26 coals.

4.2.2. Trace Elements

Compared to the average figures for world hard coals [53], the trace elements Cs (CC, 3.58), Pb (CC, 2.05) and Zr (CC, 2.04) are slightly enriched in No. 26 coals. Cr, Cd, Ba, Sr, Cu and Bi are depleted (Table 4, Figure 5A).

By comparison with the average figures for world clays [54], the roof sample is slightly enriched in Er (CC, 2.92), Pb (CC, 2.78), Bi (CC, 2.37), Yb (CC, 2.32), Lu (CC, 2.23), Ho (CC, 2.11), and Dy (CC, 2.08). Other elements are normal or depleted (Figure 5B). In floor sample, only Pb (CC, 2.23) and Bi (CC, 2.19) are slightly enriched, and other elements are normal or depleted (Figure 5C).

In addition, from the vertical section of No. 26 coal seams, the contents of Cs, Pb, and Zr in coal near the host rocks (roof and floor) outclass those in the middle coal seams (Figure 6).

4.2.3. Rare Earth Elements and Yttrium (REY)

REY geochemical classification adopted in our study is according to [55]. Previous studies have shown that because of interference of BaO or BaOH, the Eu content measured in coal by ICP-MS based on quadrupole should be carefully used to identify the positive Eu anomaly [56,57]. In our study, the Ba/Eu values (56 to 370, average of 185), and the relation degree of Ba and Eu in coal samples are low (Figure 7), indicating that Ba concentration has no interference with Eu content.

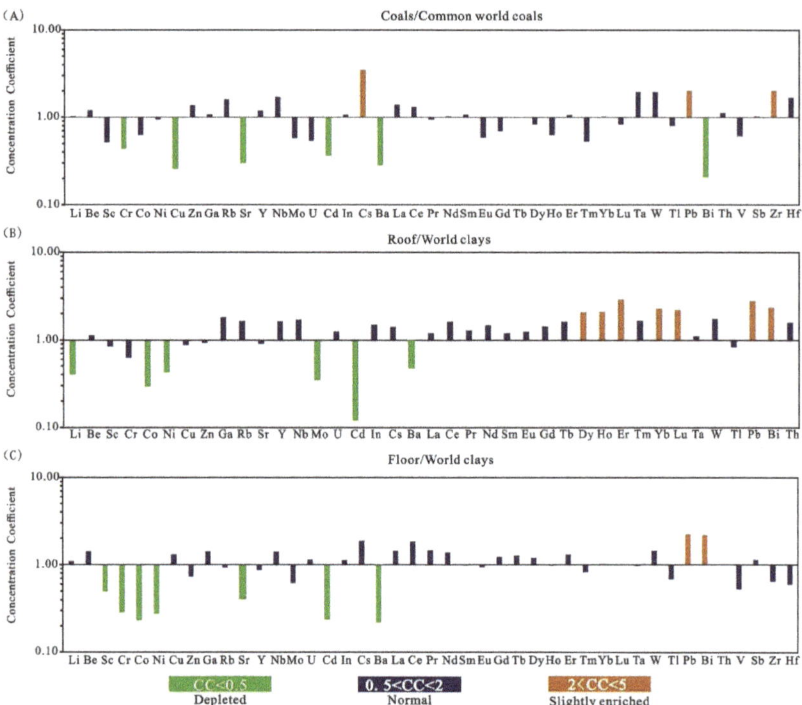

Figure 5. CC [48] of trace elements in (**A**) coals, (**B**) the roof, and (**C**) the floor in the Junde mine. Average concentration of trace elements of world hard coals is used to normalize the corresponding element concentration in studied coal samples [53]. Average concentration of trace element of world clays is used to normalize the corresponding element concentration in the studied roof and floor samples [54].

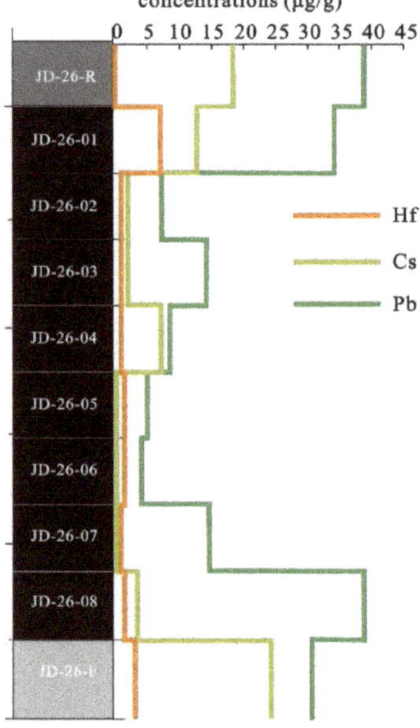

Figure 6. Variations of Hf, Cs, and Pb contents in the Junde mine.

Table 4. Concentrations of trace elements in Junde mine samples, on a whole coal basis (μg/g).

Sample	Li	Be	Sc	Cr	Co	Ni	Cu	Zn	Ga	Rb	Sr	Y	Nb	Mo	U	Cd	In	Cs	Ba	La	Ce
JD-26-R	22	3.4	13	69	5.5	20	32	83	29	221	217	51	19	0.56	5.4	0.11	0.09	18	221	58	122
JD-26-01	19	4.3	4.1	4.9	7.1	11	3.4	32	11	65	52	29	11	1.3	2.0	0.14	0.08	13	51	28	52
JD-26-02	7.5	3	1.5	5	4.3	10	4.5	16	4.3	16	42	6.7	4.6	0.83	0.89	0.04	0.02	2.0	47	6.4	14
JD-26-03	16	1.7	1.4	6.6	1.4	9	5.4	21	6.4	16	35	6.3	4.5	1.3	0.98	0.03	0.02	2.0	66	9.9	22
JD-26-04	23	1.6	2.9	14	1.7	16	2.8	43	9.3	55	27	12	6.8	1.1	2.2	0.15	0.03	7.2	35	42	72
JD-26-05	4.8	1.1	1.1	4.5	2.4	21	2.4	22	4.2	5.9	23	6.5	4.7	0.91	0.86	0.06	0.03	0.45	28	8.5	17
JD-26-06	4.6	0.62	0.87	4.8	1.7	5.7	3.6	57	4.1	5.1	27	4.8	8.1	1.6	0.56	0.11	0.02	0.58	27	6.0	12
JD-26-07	8.1	1.3	1.7	5.9	2.8	10	3.7	19	4.5	4.4	31	6.7	3.9	1.8	1.1	0.05	0.02	0.42	34	7.8	19
JD-26-08	17	1.9	2.6	10	4.1	14	6.2	43	6.7	17	27	9.0	6.8	1.5	1.7	0.06	0.05	3.4	55	15	34
JD-26-F	59	4.3	7.4	31	4.4	13	48	65	23	124	97	27	16	0.99	4.9	0.22	0.07	24	104	69	138
Average- coal	13	1.9	2.0	7.0	3.2	12	4.0	32	6.3	23	33	10	6.3	1.3	1.3	0.08	0.03	3.6	43	15	30
world coal [a]	12	1.6	3.9	16	5.1	13	16	23	5.8	14	110	8.4	3.7	2.2	2.4	0.22	0.03	1	150	11	23
world clay [b]	54	3	15	110	19	49	36	89	16	133	240	31	11	1.6	4.3	0.91	0.06	13	460	48	75

Sample	Pr	Nd	Sm	Eu	Gd	Tb	Dy	Ho	Er	Tm	Yb	Lu	Ta	Tl	Pb	Bi	Th	V	Sb	Zr	Hf
JD-26-R	13	53	9.6	1.5	8.3	1.4	9.2	1.9	5.6	0.84	5.8	0.87	1.6	1.1	39	0.9	22	-	-	-	-
JD-26-01	5.9	23	4.1	0.40	3.6	0.68	4.2	0.86	2.5	0.42	2.6	0.441	1.7	0.496	34	0.244	9.3	12	2.2	205	7.0
JD-26-02	1.5	5.9	1.1	0.18	1.0	0.19	1.1	0.21	0.59	0.11	0.74	0.11	0.26	0.29	7.0	0.08	2.2	14.8	1.8	40	0.98
JD-26-03	2.4	8.6	1.6	0.18	1.3	0.23	1.1	0.23	0.73	0.11	0.78	0.13	0.42	0.35	14	0.16	3.2	16.5	0.39	46	1.3
JD-26-04	7.7	26	4.1	0.62	3.8	0.53	2.6	0.41	1.3	0.20	1.2	0.168	0.53	0.57	8.5	0.33	3.9	20.9	0.57	44	1.3
JD-26-05	2.0	7.2	1.3	0.16	1.1	0.20	1.1	0.23	0.65	0.12	0.70	0.11	0.26	0.25	4.9	0.07	2.5	5.7	0.31	66	1.5
JD-26-06	1.4	5.4	1.0	0.11	0.84	0.16	0.88	0.17	0.50	0.08	0.48	0.08	0.61	0.32	4.3	0.07	1.6	9.5	0.48	70	1.5
JD-26-07	2.2	9.0	1.7	0.22	1.4	0.24	1.2	0.25	0.77	0.12	0.69	0.13	0.24	0.54	14.6	0.23	2.4	13	0.90	47	1.1
JD-26-08	3.5	14	2.5	0.35	2.1	0.34	1.9	0.35	0.94	0.16	1.0	0.17	0.48	1.2	39	0.43	4.7	29	1.0	72	1.8
JD-26-F	14.6	50	7.9	1.1	7.2	1.1	5.3	0.89	2.5	0.41	2.6	0.39	1.4	0.89	31	0.83	14	64	1.5	122	3.0
Average-coal	3.3	12	2.2	0.28	1.9	0.32	1.7	0.34	0.99	0.16	1.0	0.17	0.55	0.5	16	0.2	3.7	15	0.95	73	2.0
world coal [a]	3.5	12	2	0.47	2.7	0.32	2.1	0.54	0.93	0.31	1.0	0.2	0.28	0.63	7.8	0.97	3.3	25	0.92	36	1.2
world clay [b]	10	36	8	1.2	5.8	0.83	4.4	0.9	1.9	0.5	2.5	0.39	1.4	1.3	14	0.38	14	120	1.3	190	5

[a], averages of world coals [53]; [b], averages of world clay [54]; -, no test or no result.

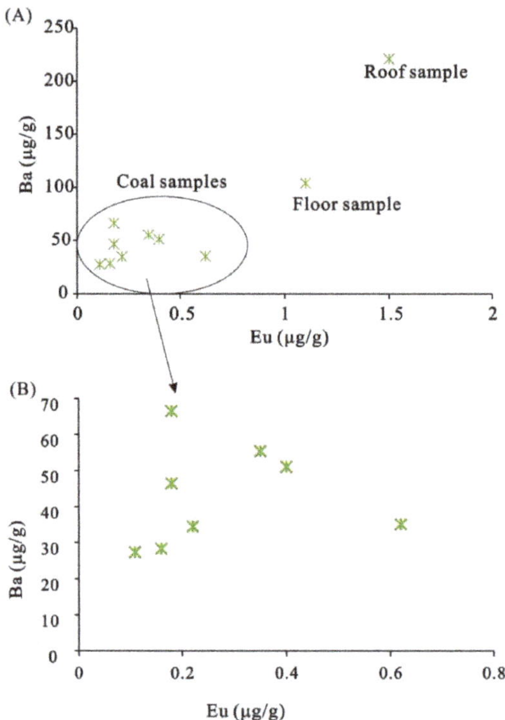

Figure 7. (**A**) Correlation diagram of Eu and Ba contents in coals, roof, and floor samples. (**B**) Magnification of relationship between Eu and Ba in coal samples from Junde mine.

The contents of REY of No. 26 coal in the Junde mine (from 34.1 µg/g to 174.6 µg/g, average 80.6 µg/g; Table 5) slightly exceed those in the world's hard coal [53], but below those in Chinese coals [22] and the upper continental crust [58].

Table 5. REY geochemical parameters for the studied samples from the Junde mine.

Sample	LREY	MREY	HREY	REY	La_N/Lu_N	La_N/Sm_N	Gd_N/Lu_N	Eu_N/Eu_N *	Ce_N/Ce_N *	δY
JD-26-R	255.1	70.9	15.0	340.9	0.99	1.07	0.73	0.72	1.17	1.33
JD-26-01	112.6	37.5	6.8	157.0	0.96	1.24	0.63	0.43	1.06	1.67
JD-26-02	28.6	9.2	1.8	39.6	0.90	1.01	0.74	0.66	1.15	1.57
JD-26-03	44.4	9.0	2.0	55.4	1.13	1.14	0.73	0.53	1.19	1.35
JD-26-04	151.9	19.5	3.3	174.6	3.72	1.80	1.73	0.70	1.05	1.47
JD-26-05	36.1	9.0	1.8	46.9	1.16	1.18	0.75	0.55	1.11	1.40
JD-26-06	26.0	6.8	1.3	34.1	1.09	1.05	0.78	0.48	1.11	1.39
JD-26-07	39.6	9.7	2.0	51.2	0.92	0.82	0.81	0.57	1.21	1.32
JD-26-08	69.7	13.7	2.7	86.0	1.34	1.09	0.92	0.64	1.23	1.27
JD-26-F	279.5	41.6	6.7	327.8	2.67	1.57	1.40	0.66	1.14	1.52

Notes: Units for REYs: µg/g; REY = LREY + MREY + HREY; Calculation formula for LREY, MREY, HREY, Eu_N/Eu_N *, Ce_N/Ce_N *, δY are according to [56].

The REY contents in host rocks (roof and floor) are about four times that of coals (Table 5). Additionally, REY contents of the roof (340.9 µg/g) and the floor (327.8 µg/g) are well above those of average world clays [54] and the upper continental crust [58].

According to [55], the features of REY in No. 26 coals from Junde mine are LREY enrichment, with obviously negative Eu (Figure 8A), and, in the upper coal bench only, by HREY enrichment with obviously negative Eu, and positive Y anomalies. Correspondingly, the feature of REY of the roof is HREY enrichment like the top coal sample, while the feature of REY of the floor is LREY enrichment, like other coal samples (Figure 8B).

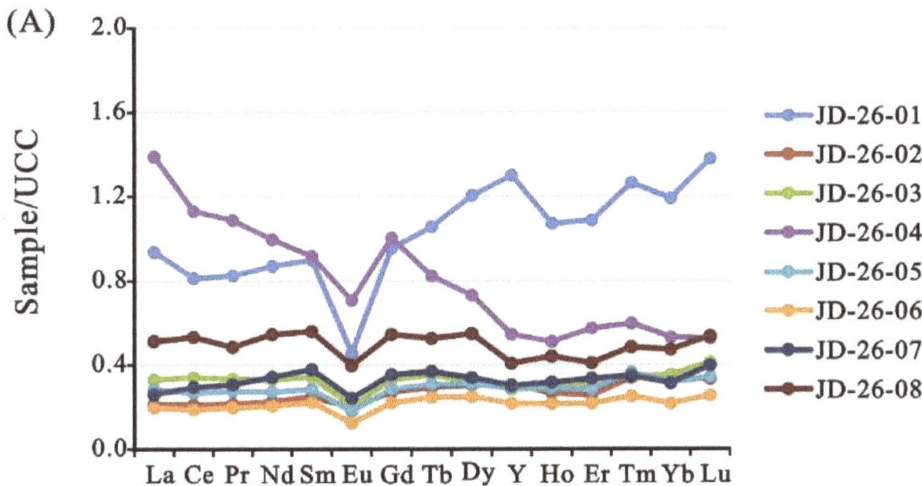

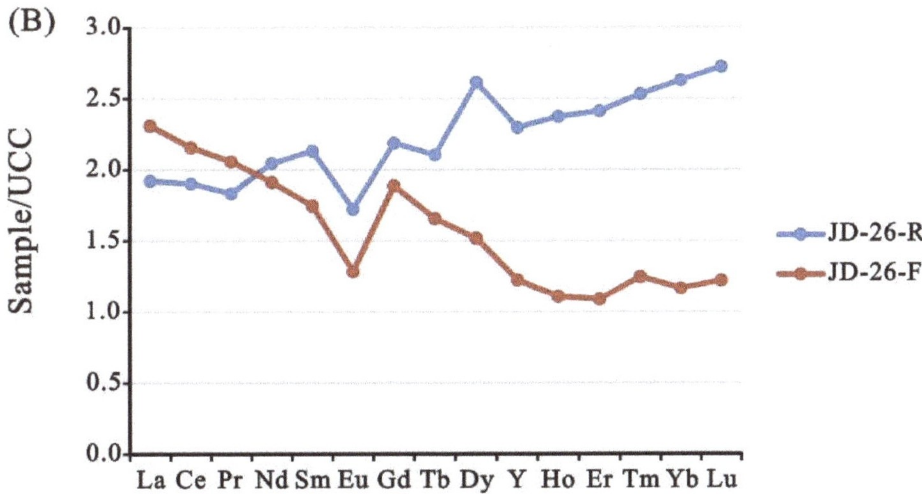

Figure 8. REY characteristics for (**A**) coal samples, (**B**) roof and floor samples in Junde mine. REY contents were normalized by UCC.

4.3. Mineralogy

Based on X-ray diffraction experiment (XRD) results (Figure 9), clay minerals and quartz are the chief minerals in the Junde coals, and after them calcite, siderite, and pyrite. Some monazite and zircon were detected in No. 26 coal by SEM-EDS. The minerals in the floor and roof are primarily clay minerals, quartz, potassium feldspar, and plagioclase.

4.3.1. Clay Minerals

Clay minerals in No. 26 coals primarily occur as massive lumps (Figure 10A,B) and cell-fillings (Figure 3C), indicating authigenic and terrigenous origins. Kaolinite, illite/smectite formation (I/S), and illite have been detected in Junde coals.

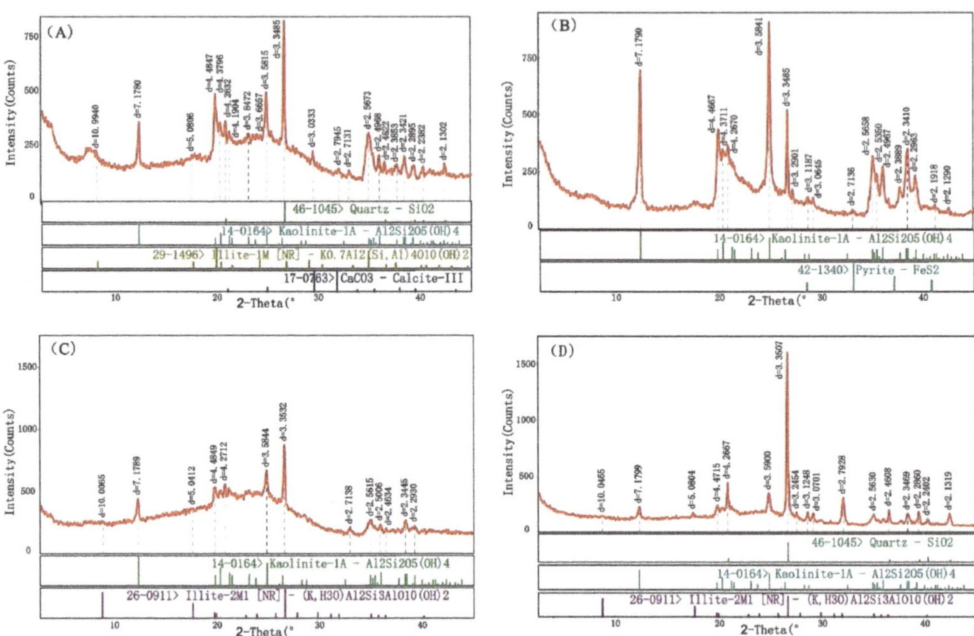

Figure 9. XRD patterns of No. 26 coals of Junde mine. (**A**) Sample JD-26-02; (**B**) sample JD-26-04; (**C**) sample JD-26-05; (**D**) sample JD-26-08.

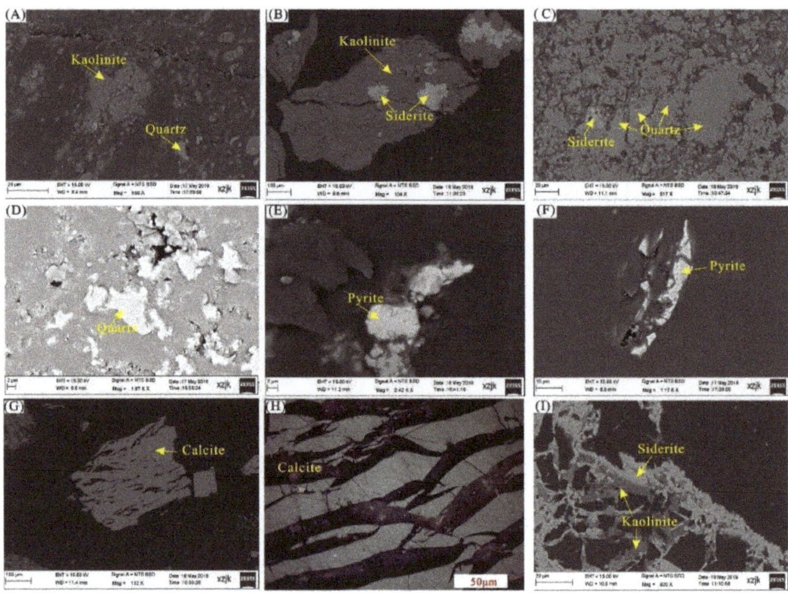

Figure 10. Minerals in samples from Junde mine (SEM, SE mode). (**A**) Kaolinite and quartz in sample JD-26-05; (**B**) Kaolinite and siderite in sample JD-26-06; (**C**) Siderite and quartz in sample JD-26-04; (**D**) Quartz in sample JD-26-05; (**E**) Pyrite in sample JD-26-04; (**F**) Pyrite in sample JD-26-05; (**G**) Calcite in sample JD-26-04; (**H**) Fracture-filling calcite in sample JD-26-05, Reflected light, oil immersion; (**I**) Siderite and kaolinite in sample JD-26-06.

4.3.2. Quartz

In No. 26 coals, some quartz occurs as fine-grained particles (Figure 10C,D), indicating a syngenetic detrital origin. A lot of quartz in the studied coals has sharp edges Figures 3B,D and 10A) and indicates the influence of high temperature and volcanic activity [35,59,60].

4.3.3. Pyrite

Some pyrite in No. 26 coal is in the form of discrete crystals (Figures 3B,D and 10E), suggesting an authigenic origin. In addition, fracture fillings pyrite has also been found (Figures 3A,G and 10F), suggestive of an epigenetic origin [61].

4.3.4. Carbonate Minerals

Calcite and siderite are the major carbonate minerals. Calcite generally occurs as plates (Figure 10G) and crack fillings (Figure 10H), suggesting an authigenic origin and epigenetic origin, respectively. Additionally, the siderite occurs as fracture-fillings cutting through the kaolinite (Figure 10I), indicating an epigenetic origin.

4.3.5. Other Minerals

Zircon (Figure 11) and monazite (Figure 12) were observed in No. 26 coal. Zircon occurs as corroded-crystal (Figure 11A), hexagon (Figure 11B) or incomplete quadrilateral bipyramid (Figure 11C), while monazite occurs mainly as nearly circular individual particles (Figure 12).

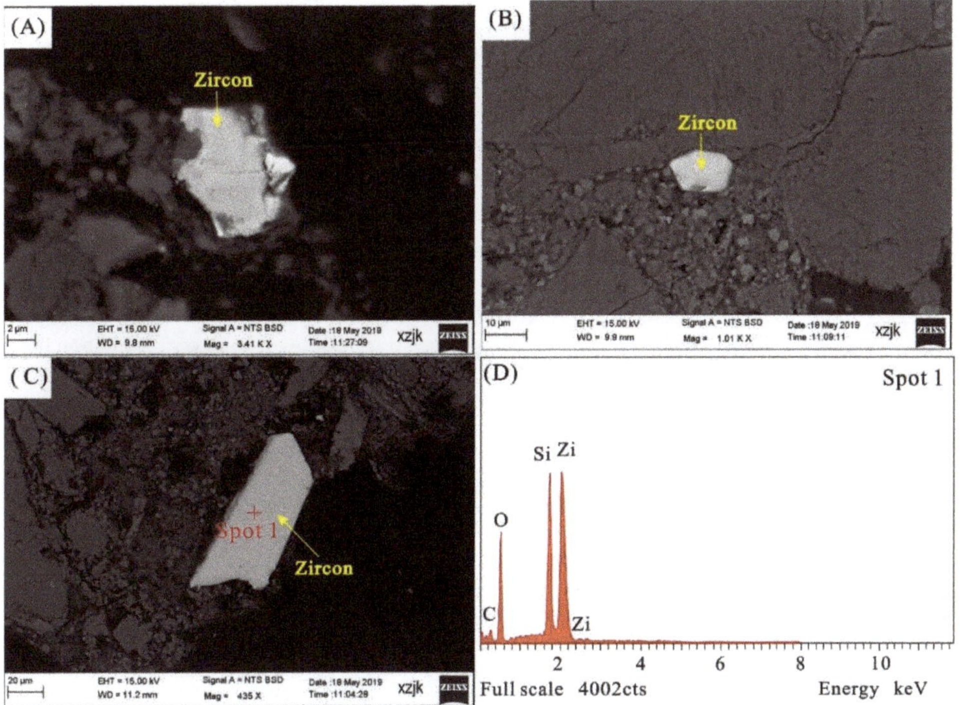

Figure 11. Zircon in samples from Junde mine. (**A**) Sample JD-26-03; (**B**) Sample JD-26-05; (**C**) Sample JD-26-06. (**D**) EDS spectrum of Spot 1 in Figure 11C.

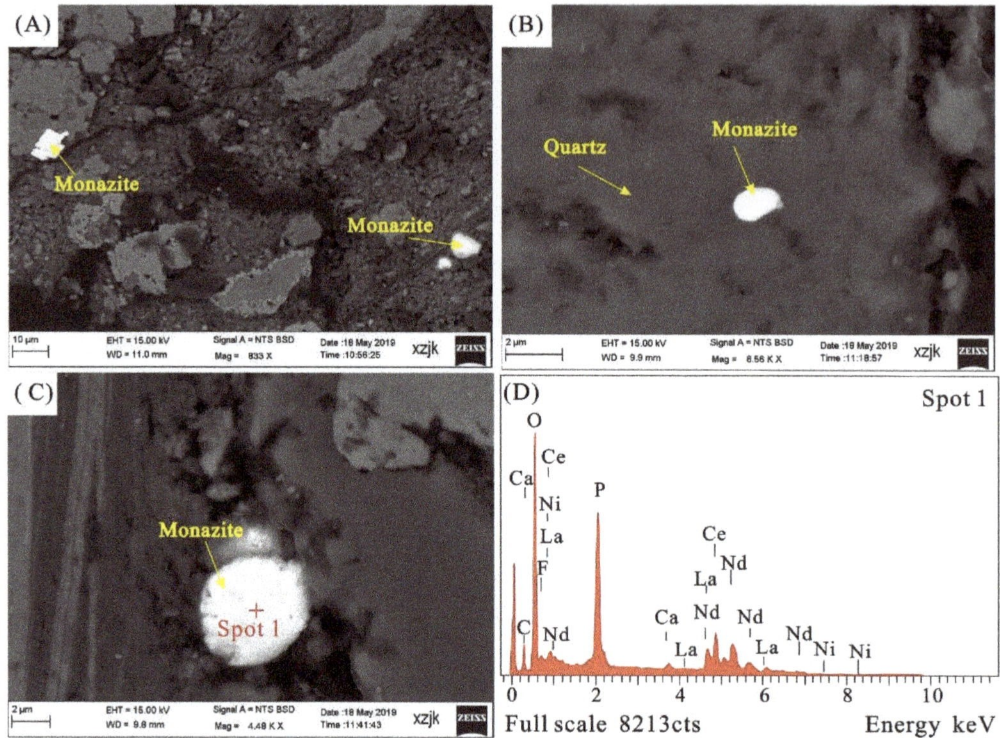

Figure 12. Monazite in samples from Junde mine. (**A**) Sample JD-26-04; (**B**) Sample JD-26-06; (**C**) Sample JD-26-07; (**D**) EDS spectrum of Spot 1 in Figure 12C.

5. Discussion

5.1. Sediment Source

The value of Al_2O_3/TiO_2 is commonly used to infer the sediment source (including coal deposits) [62,63]. The value of Al_2O_3/TiO_2 in Junde coals varies from 26 to 45 (average 22.6) (Figure 13A), suggesting source rocks during No. 26 coal formation in Junde mine are primarily felsic rocks. This conclusion can also be evidenced by the diagrams of Zr/Sc—Th/Sc (Figure 13B), which have been proved a useful provenance indicator for coal deposits [64,65].

The No. 26 coals in the study area and the Fuqiang coals from the Hunchun Coalfield [33] have similar Cs-Pb-Zr enrichment characteristics. The provenance in Fuqiang coals is Mesozoic and Paleozoic igneous and metamorphic rocks [33]. To further explore the terrigenous supply of No. 26 coal, a comparison about Sr/Y, La/Yb, and Al_2O_3/TiO_2 values of Junde coals with the local (the Jiamusi block) rocks of Neoproterozoic, Late Paleozoic, Mesozoic and Cretaceous ages [66–69] is shown in Figure 14. Figure 14 indicates that the source material of the Junde mine mainly comes from the Late Paleozoic igneous rocks from Jiamusi block. In addition, the REY patterns of No. 26 coal are akin to those of the Late Paleozoic igneous rocks with negative Eu anomalies (Figure 15). Sun et al. have also determined that detrital zircons of Mesozoic and Paleozoic ages predominate in sandstones of the Chengzihe Formation [70]. Those zircons are shaped similarly to the zircons described in this study (Figure 11A–C). Therefore, the terrigenous components in No. 26 coal from Junde mine were mainly derived from the Late Paleozoic meta-igneous rocks in local block.

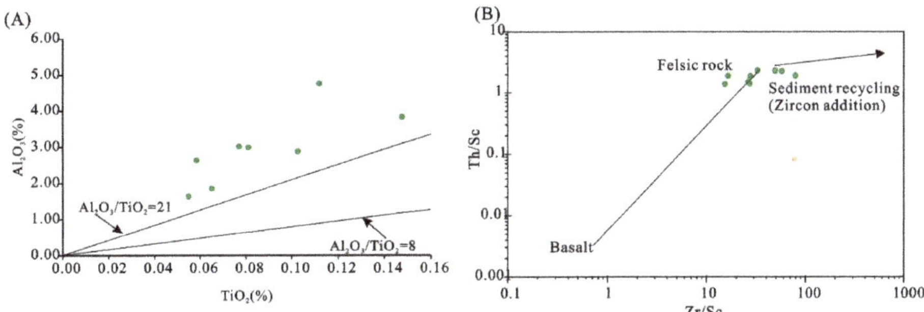

Figure 13. Plot of (**A**) TiO_2 vs. Al_2O_3 and (**B**) Zr/Sc vs. Th/Sc for coal samples from the Junde mine.

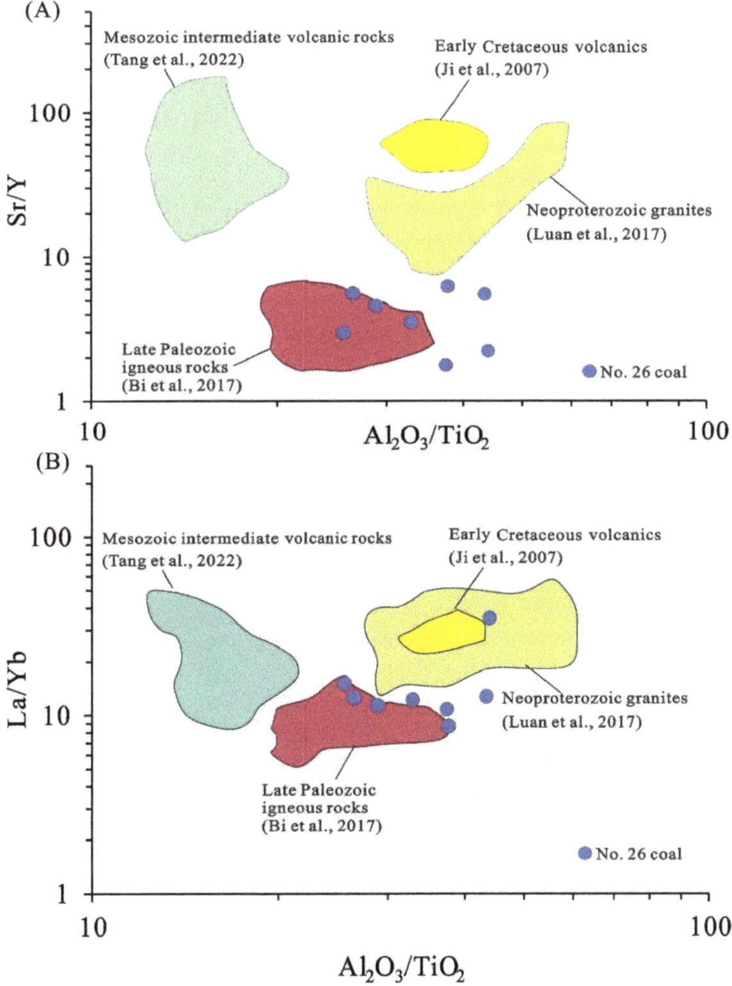

Figure 14. Comparison of (**A**) Al_2O_3/TiO_2 vs. Sr/Y and (**B**) Al_2O_3/TiO_2 vs. La/Yb between the studied samples, and the local (the Jiamusi block) rocks of Neoproterozoic, Late Paleozoic, Mesozoic and Cretaceous ages [66–69].

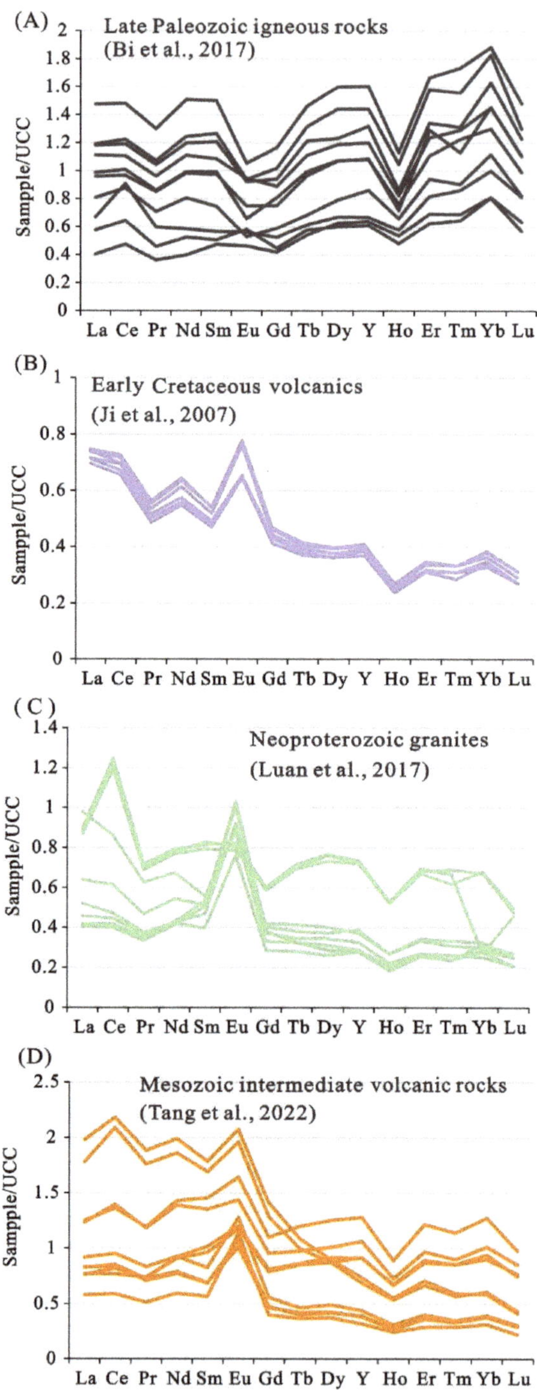

Figure 15. Characteristics of REY of (**A**) Late Paleozoic rocks, (**B**) Early Cretaceous rocks, (**C**) the Neoproterozoic rocks, and (**D**) Mesozoic rocks in the Jiamusi block [66–69].

5.2. Sedimentary Environment

The value of Sr/Ba has been widely used to identify the depositional environment of sedimentary rocks and coals [71,72]. Generally, coals with Sr/Ba ratio greater than 1 and less than 1 was formed in the environment of seawater and freshwater intrusion, respectively. However, we should be cautious when using this indicator, because the content of Sr and Ba in terrigenous clastic minerals (especially feldspar) may cause misjudgment of the sedimentary environment [73,74]. In the present study, feldspar content is so low that it is not detected by XRD. Thus, Sr/Ba ratio is used to roughly distinguish sedimentary environment in this study. Sr/Ba value of No. 26 coal seam from Junde mine varies from 0.48 to 1.01, and the average Sr/Ba value of the coal seam is 0.81, suggestive of a primarily fresh-water affected process of No. 26 coal.

Because NE China is abundant with Cu porphyry ores [75] that may be eroded and consequently changing the ratio in terrigenous sediments, Sr/Cu ratio is not suitable to infer the sedimentary environment. The relationship of sulfur and the environment has been widely recognized [72]. In the Junde mine, sulfate sulfur increases with depth (Figure 16), which indicate oxidation/evaporation gradually decreases during No.26 coal formation.

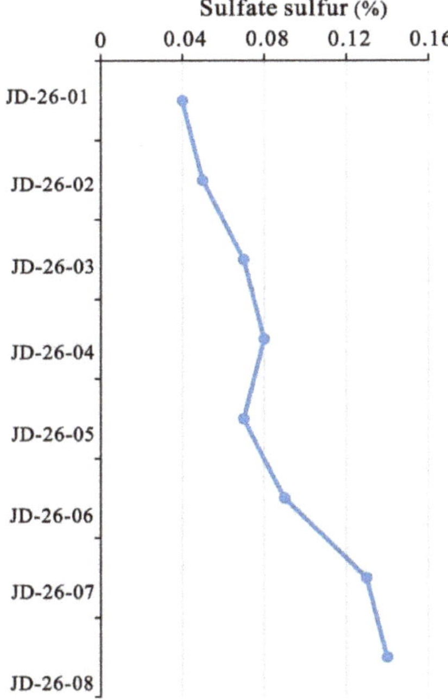

Figure 16. Regularity of the variation of sulfate sulfur in the Junde samples.

5.3. Influence of Volcanic Ash

It is generally believed that the Eu anomaly in coal does not originate from the weathering process during the transportation of metals from provenance to coal forming peat, but is inherited from rocks in the source areas [56]. Eu anomalies characteristics of No. 26 coal from Junde mine (including roof and floor) resembles those of felsic volcanic rocks.

Previous studies suggested that the crystalline habit and morphology of terrigenous detrital zircons are quite different from those of igneous detrital zircons [76]. The characteristics of the former are tetragonal bipyramids with relatively short prisms, and the length width ratio (c/a values) is about two [77], while the feature of the latter is long and well-developed, with a c/a value of more than 2.5 [77]. In the present study, the ratio of c/a in the zircon found in No. 26 coal (Figure 11C) is > 2.5, indictive of a pyroclastic origin. Another possibility that its angular and elongated shapes may evidence just short transportation by relatively quiet water streams cannot be ruled out.

A large number of sharp-edged quartz particles were observed in No. 26 coal (Figures 3B,D and 10A), indicating a volcanic origin rather than a terrigenous clastic origin [35,59,60].

Although zircon and some high-T quartzs have been found in the No. 26 coals, the vermicular kaolinite/chlorite particles that commonly indicate volcanic ash have not been found. Thus, the volcanic ash contribution to No. 26 coal seems very low, if occurred. This conclusion is consistent with [70].

6. Conclusions

The characteristics of minerals and elements of No. 26 coal from Junde mine, Hegang coalfield, northeastern China, were studied. The main conclusions are summarized as follows.

(1) The No. 26 coal of Junde mine is slightly enriched in Cs, Pb, and Zr. From the vertical section of No. 26 coal seams, the contents of Cs, Pb, and Zr in the coal samples near the host rocks (roof and floor) outclass those in the middle coal seams.

(2) The minerals in Junde coals mainly include clay minerals and quartz, followed by calcite, siderite, pyrite, monazite, and zircon.

(3) The terrigenous components in No. 26 coal from Junde mine were derived from the Late Paleozoic meta-igneous rocks in the Jiamusi block evidenced by Sr/Y, La/Yb, and Al_2O_3/TiO_2 ratios, Zr/Sc—Th/Sc plot, and negative Eu anomalies.

(4) The No. 26 coal from the Junde mine was affected by fresh water during coal formation. Sulfate sulfur indicate oxidation/evaporation gradually decreases during No. 26 coal formation.

(5) The volcanic ash contribution to No. 26 coal seems very low, if occurred.

Author Contributions: Y.W.: Methodology, Data curation, Writing-original draft. W.H.: Data curation, Writing-original draft. G.Q.: Methodology, Data curation, Writing-original draft. A.W.: Supervision. D.C.: Conceptualization, Supervision. All authors have read and agreed to the published version of the manuscript.

Funding: This work was funded by the National Key Research and Development Plan of China (2021YFC2902004), National Natural Science Foundation of China (41972174 and 42072197), and Science Foundation of Hebei Normal University (L2021B25).

Institutional Review Board Statement: Not applicable.

Informed Consent Statement: Not applicable.

Data Availability Statement: Available from Yingchun Wei and corresponding author.

Conflicts of Interest: The authors declare no conflict of interest.

References

1. Finkelman, R.B.; Dai, S.; French, D. The importance of minerals in coal as the hosts of chemical elements: A review. *Int. J. Coal Geol.* **2019**, *212*, 103251. [CrossRef]
2. Dai, S.; Finkelman, R.B. Coal as a promising source of critical elements: Progress and future prospects. *Int. J. Coal Geol.* **2018**, *186*, 155–164. [CrossRef]
3. Seredin, V.V.; Dai, S.; Sun, Y.; Chekryzhov, I.Y. Coal deposits as promising sources of rare metals for alternative power and energy-efficient technologies. *Appl. Geochem.* **2013**, *31*, 1–11. [CrossRef]
4. Dai, S.; Finkelman, R.B.; French, D.; Hower, J.C.; Graham, I.T.; Zhao, F. Modes of occurrence of elements in coal: A critical evaluation. *Earth-Sci. Rev.* **2021**, *222*, 103815. [CrossRef]
5. Liu, J.; Dai, S.; Song, H.; Nechaev, V.P.; French, D.; Spiro, B.F.; Graham, I.T.; Hower, J.C.; Shao, L.; Zhao, J. Geological factors controlling variations in the mineralogical and elemental compositions of Late Permian coals from the Zhijin-Nayong Coalfield, western Guizhou, China. *Int. J. Coal Geol.* **2021**, *247*, 103855. [CrossRef]
6. Dai, S.; Chekryzhov, I.Y.; Seredin, V.V.; Nechaev, V.P.; Graham, I.T.; Hower, J.C.; Ward, C.R.; Ren, D.; Wang, X. Metalliferous coal deposits in East Asia (Primorye of Russia and South China): A review of geodynamic controls and styles of mineralization. *Gondwana Res.* **2016**, *29*, 60–82. [CrossRef]

7.   Seredin, V.V.; Finkelman, R.B. Metalliferous coals: A review of the main genetic and geochemical types. *Int. J. Coal Geol.* **2008**, *76*, 253–289. [CrossRef]

8.   Chehreh Chelgani, S.; Hower, J.C. Estimating REY content of eastern Kentucky coal samples based on their associated ash elements. *J. Rare Earths* **2018**, *36*, 1234–1238. [CrossRef]

9.   Taggart, R.K.; Hower, J.C.; Hsu-Kim, H. Effects of roasting additives and leaching parameters on the extraction of rare earth elements from coal fly ash. *Int. J. Coal Geol.* **2018**, *196*, 106–114. [CrossRef]

10.   Kolker, A.; Hower, J.C.; Karamalidis, A.K. Introduction to critical elements in coal and coal ash and their recovery, a virtual special issue. *Int. J. Coal Geol.* **2019**, *206*, 19–20. [CrossRef]

11.   Arbuzov, S.; Chekryzhov, I.; Finkelman, R.; Sun, Y.; Zhao, C.; Il'Enok, S.; Blokhin, M.; Zarubina, N. Comments on the geochemistry of rare-earth elements (La, Ce, Sm, Eu, Tb, Yb, Lu) with examples from coals of north Asia (Siberia, Russian far East, North China, Mongolia, and Kazakhstan). *Int. J. Coal Geol.* **2019**, *206*, 106–120. [CrossRef]

12.   Bagdonas, D.; Enriquez, A.; Coddington, K.; Finnoff, D.; McLaughlin, J.; Bazilian, M.; Phillips, E.; McLing, T. Rare earth element resource evaluation of coal byproducts: A case study from the Powder River Basin, Wyoming. *Renew. Sustain. Energy Rev.* **2022**, *158*, 112148. [CrossRef]

13.   Zhou, M.; Zhao, L.; Wang, X.; Nechaev, V.P.; French, D.; Spiro, B.F.; Graham, I.T.; Hower, J.C.; Dai, S. Mineralogy and geochemistry of the Late Triassic coal from the Caotang mine, northeastern Sichuan Basin, China, with emphasis on the enrichment of the critical element lithium. *Ore Geol. Rev.* **2021**, *139*, 104582. [CrossRef]

14.   Hu, S.; Ni, Y.; Yin, Q.; Wang, J.; Lv, L.; Cen, K.; Zhou, H. Research on element migration and ash deposition characteristics of high-alkali coal in horizontal liquid slagging cyclone furnace. *Fuel* **2022**, *308*, 121962. [CrossRef]

15.   Sutcu, E.C.; Şentürk, S.; Kapıcı, K.; Gökçe, N. Mineral and rare earth element distribution in the Tunçbilek coal seam, Kütahya, Turkey. *Int. J. Coal Geol.* **2021**, *245*, 103820. [CrossRef]

16.   Hower, J.C.; Eble, C.F.; Xie, P.; Liu, J.; Fu, B.; Hood, M.M. Aspects of rare earth element enrichment in Allegheny Plateau coals, Pennsylvania, USA. *Appl. Geochem.* **2022**, *136*, 105150. [CrossRef]

17.   Kolker, A.; Scott, C.; Lefticariu, L.; Mastalerz, M.; Drobniak, A.; Scott, A. Trace element partitioning during coal preparation: Insights from U.S. Illinois Basin coals. *Int. J. Coal Geol.* **2021**, *243*, 103781. [CrossRef]

18.   Lei, Z.; Cheng, Z.; Ling, Q.; Liu, X.; Cui, P.; Zhao, Z. Investigating the trigger mechanism of Shenfu bituminous coal pyrolysis. *Fuel* **2022**, *313*, 122995. [CrossRef]

19.   Wang, X.; Bian, J.; Zeng, F.; Pan, Z.; Chai, P. Aluminum-bearing nano-sized minerals in vitrain band and its implications for modes of occurrence of Al in Carboniferous-Permian coals from the Hedong Coalfield, northern China. *Int. J. Coal Geol.* **2021**, *248*, 103861. [CrossRef]

20.   BaiLin, Y.; XiaoFeng, J. An Investigation on Configuration status of Science and Technology Resource in the Coal Industry of Heilongjiang Province. *Energy Procedia* **2011**, *5*, 2167–2171. [CrossRef]

21.   Zhang, Y.; Sun, Y.; Qin, J. Sustainable development of coal cities in Heilongjiang province based on AHP method. *Int. J. Min. Sci. Technol.* **2012**, *22*, 133–137. [CrossRef]

22.   Dai, S.; Ren, D.; Chou, C.-L.; Finkelman, R.B.; Seredin, V.V.; Zhou, Y. Geochemistry of trace elements in Chinese coals: A review of abundances, genetic types, impacts on human health, and industrial utilization. *Int. J. Coal Geol.* **2012**, *94*, 3–21. [CrossRef]

23.   Zhao, L.; Ward, C.R.; French, D.; Graham, I.T. Mineralogical composition of Late Permian coal seams in the Songzao Coalfield, southwestern China. *Int. J. Coal Geol.* **2013**, *116–117*, 208–226. [CrossRef]

24.   Zhuang, X.; Querol, X.; Alastuey, A.; Plana, F.; Moreno, N.; Andrés, J.M.; Wang, J. Mineralogy and geochemistry of the coals from the Chongqing and Southeast Hubei coal mining districts, South China. *Int. J. Coal Geol.* **2007**, *71*, 263–275. [CrossRef]

25.   Jiu, B.; Huang, W.; Mu, N. Mineralogy and elemental geochemistry of Permo-Carboniferous Li-enriched coal in the southern Ordos Basin, China: Implications for modes of occurrence, controlling factors and sources of Li in coal. *Ore Geol. Rev.* **2022**, *141*, 104686. [CrossRef]

26.   Pan, J.; Ge, T.; Liu, W.; Wang, K.; Wang, X.; Mou, P.; Wu, W.; Niu, Y. Organic matter provenance and accumulation of transitional facies coal and mudstone in Yangquan, China: Insights from petrology and geochemistry. *J. Nat. Gas Sci. Eng.* **2021**, *94*, 104076. [CrossRef]

27.   Wei, Y.; He, W.; Qin, G.; Fan, M.; Cao, D. Lithium Enrichment in the No. 21 Coal of the Hebi No. 6 Mine, Anhe Coalfield, Henan Province, China. *Minerals* **2020**, *10*, 521. [CrossRef]

28.   Qin, G.; Cao, D.; Wei, Y.; Wang, A.; Liu, J. Geochemical characteristics of the Permian coals in the Junger-Hebaopian mining district, northeastern Ordos Basin, China: Key role of paleopeat-forming environments in Ga-Li-REY enrichment. *J. Geochem. Explor.* **2020**, *213*, 106494. [CrossRef]

29.   Qin, G.; Cao, D.; Wei, Y.; Wang, A.; Liu, J. Mineralogy and Geochemistry of the No. 5-2 High-Sulfur Coal from the Dongpo Mine, Weibei Coalfield, Shaanxi, North China, with Emphasis on Anomalies of Gallium and Lithium. *Minerals* **2019**, *9*, 402. [CrossRef]

30.   Zhuang, X.; Querol, X.; Alastuey, A.; Juan, R.; Plana, F.; Lopez-Soler, A.; Du, G.; Martynov, V.V. Geochemistry and mineralogy of the Cretaceous Wulantuga high-germanium coal deposit in Shengli coal field, Inner Mongolia, Northeastern China. *Int. J. Coal Geol.* **2006**, *66*, 119–136. [CrossRef]

31.   Qi, H.; Hu, R.; Zhang, Q. REE Geochemistry of the Cretaceous lignite from Wulantuga Germanium Deposit, Inner Mongolia, Northeastern China. *Int. J. Coal Geol.* **2007**, *71*, 329–344. [CrossRef]

32. Shao, L.; Hou, H.; Tang, Y.; Lu, J.; Qiu, H.; Wang, X.; Zhang, J. Selection of strategic replacement areas for CBM exploration and development in China. *Nat. Gas Ind. B* **2015**, *2*, 211–221. [CrossRef]

33. Guo, W.; Dai, S.; Nechaev, V.P.; Nechaeva, E.V.; Wei, G.; Finkelman, R.B.; Spiro, B.F. Geochemistry of Palaeogene coals from the Fuqiang Mine, Hunchun Coalfield, northeastern China: Composition, provenance, and relation to the adjacent polymetallic deposits. *J. Geochem. Explor.* **2019**, *196*, 192–207. [CrossRef]

34. Guo, W.; Nechaev, V.; Yan, X.; Yang, C.; He, X.; Shan, K.; Nechaeva, E.V. New data on geology and germanium mineralization in the Hunchun Basin, northeastern China. *Ore Geol. Rev.* **2019**, *107*, 381–391. [CrossRef]

35. Dai, S.; Guo, W.; Nechaev, V.P.; French, D.; Ward, C.R.; Spiro, B.F.; Finkelman, R.B. Modes of occurrence and origin of mineral matter in the Palaeogene coal (No. 19-2) from the Hunchun Coalfield, Jilin Province, China. *Int. J. Coal Geol.* **2018**, *189*, 94–110. [CrossRef]

36. Nechaev, V.P.; Bechtel, A.; Dai, S.; Chekryzhov, I.; Pavlyutkin, B.I.; Vysotskiy, S.V.; Ignatiev, A.V.; Velivetskaya, T.A.; Guo, W.; Tarasenko, I.; et al. Bio-geochemical evolution and critical element mineralization in the Cretaceous-Cenozoic coals from the southern Far East Russia and northeastern China. *Appl. Geochem.* **2020**, *117*, 104602. [CrossRef]

37. Yang, Z.R.; Jiang, J.H. Depositional environments of the shitouhezi formation and its characteristics of the episodic coal accumulation,upper jurassic hegang basin,heilongjiang province. *Acta Sedimentol. Sin* **1997**, *15*, 58–63.

38. Li, Z.; Shao, L.Y.; Li, M.; Gao, D.; Zhen, M.Q. Analysis on the sequence stratigraphy and coal accumulation of shitouhezi formation in hegang coal basin. *J. Beijing Polytech. Coll.* **2011**, *10*, 11–17.

39. *ASTM Standard D3173-11*; Test Method for Moisture in the Analysis Sample of Coal and Coke. Annual Book of ASTM Standards. ASTM International: West Conshohocken, PA, USA, 2011.

40. *ASTM Standard D3174-11*; Test Method for Ash in the Analysis Sample of Coal and Coke. Annual Book of ASTM Standards. ASTM International: West Conshohocken, PA, USA, 2011.

41. *ASTM Standard D3175-11*; Test Method for Volatile Matter in the Analysis Sample of Coal and Coke. Annual Book of ASTM Standards. ASTM International: West Conshohocken, PA, USA, 2011.

42. International Committee for Coal and Organic Petrology (ICCP). The new inertinite classification (ICCP System 1994). *Fuel* **2001**, *80*, 459–471. [CrossRef]

43. International Committee for Coal and Organic Petrology (ICCP). The new vitrinite classification (ICCP System 1994). *Fuel* **1998**, *77*, 349–358. [CrossRef]

44. Pickel, W.; Kus, J.; Flores, D.; Kalaitzidis, S.; Christanis, K.; Cardott, B.; Misz-Kennan, M.; Rodrigues, S.; Hentschel, A.; Hamor-Vido, M.; et al. Classification of liptinite—ICCP System 1994. *Int. J. Coal Geol.* **2017**, *169*, 40–61. [CrossRef]

45. *ASTM Standard D3177-02*; Test Methods for Total Sulfur in the Analysis Sample of Coal and Coke. ASTM International: West Conshohocken, PA, USA, 2011.

46. *ASTM D2492-02*; Standard Test Method for Forms of Sulfur in Coal. ASTM International: West Conshohocken, PA, USA, 2012.

47. Dai, S.; Wang, X.; Zhou, Y.; Hower, J.C.; Li, D.; Chen, W.; Zhu, X.; Zou, J. Chemical and mineralogical compositions of silicic, mafic, and alkali tonsteins in the late Permian coals from the Songzao Coalfield, Chongqing, Southwest China. *Chem. Geol.* **2011**, *282*, 29–44. [CrossRef]

48. Dai, S.; Li, T.; Jiang, Y.; Ward, C.R.; Hower, J.C.; Sun, J.; Liu, J.; Song, H.; Wei, J.; Li, Q.; et al. Mineralogical and geochemical compositions of the Pennsylvanian coal in the Hailiushu Mine, Daqingshan Coalfield, Inner Mongolia, China: Implications of sediment-source region and acid hydrothermal solutions. *Int. J. Coal Geol.* **2015**, *137*, 92–110. [CrossRef]

49. *GB15224.1-2010*; Classification for Quality of Coal-Part 1: Ash. Standard Press of China: Beijing, China, 2010.

50. *GB15224.2-2010*; Classification for Quality of Coal-Part 2: Sulfur Content. Standard Press of China: Beijing, China, 2010.

51. Hower, J.C.; O'Keefe, J.M.K.; Wagner, N.J.; Dai, S.; Wang, X.; Xue, W. An investigation of Wulantuga coal (Cretaceous, Inner Mongolia) macerals: Paleopathology of faunal and fungal invasions into wood and the recognizable clues for their activity. *Int. J. Coal Geol.* **2013**, *114*, 44–53. [CrossRef]

52. Hower, J.C.; Misz-Keenan, M.; O'Keefe, J.M.; Mastalerz, M.; Eble, C.F.; Garrison, T.M.; Johnston, M.N.; Stucker, J. Macrinite forms in Pennsylvanian coals. *Int. J. Coal Geol.* **2013**, *116–117*, 172–181. [CrossRef]

53. Ketris, M.P.; Yudovich, Y.E. Estimations of Clarkes for Carbonaceous biolithes: World averages for trace element contents in black shales and coals. *Int. J. Coal Geol.* **2009**, *78*, 135–148. [CrossRef]

54. Grigoriev, N.A. *Chemical Element Distribution in the Upper Continental Crust*; UB RAS: Ekaterinburg, Russia, 2009.

55. Seredin, V.V.; Dai, S. Coal deposits as potential alternative sources for lanthanides and yttrium. *Int. J. Coal Geol.* **2012**, *94*, 67–93. [CrossRef]

56. Dai, S.; Graham, I.T.; Ward, C.R. A review of anomalous rare earth elements and yttrium in coal. *Int. J. Coal Geol.* **2016**, *159*, 82–95. [CrossRef]

57. Yan, X.; Dai, S.; Graham, I.T.; He, X.; Shan, K.; Liu, X. Determination of Eu concentrations in coal, fly ash and sedimentary rocks using a cation exchange resin and inductively coupled plasma mass spectrometry (ICP-MS). *Int. J. Coal Geol.* **2018**, *191*, 152–156. [CrossRef]

58. Taylor, S.R.; McLennan, S.M. The geochemical evolution of the continental crust. *Rev. Geophys.* **1995**, *33*, 241–265. [CrossRef]

59. Dai, S.; Ward, C.R.; Graham, I.T.; French, D.; Hower, J.C.; Zhao, L.; Wang, X. Altered volcanic ashes in coal and coal-bearing sequences: A review of their nature and significance. *Earth-Sci. Rev.* **2017**, *175*, 44–74. [CrossRef]

60. Dai, S.; Ren, D.; Hou, X.; Shao, L. Geochemical and mineralogical anomalies of the late Permian coal in the Zhijin coalfield of southwest China and their volcanic origin. *Int. J. Coal Geol.* **2003**, *55*, 117–138. [CrossRef]

61. Li, J.; Zhuang, X.; Querol, X.; Font, O.; Izquierdo, M.; Wang, Z. New data on mineralogy and geochemistry of high-Ge coals in the Yimin coalfield, Inner Mongolia, China. *Int. J. Coal Geol.* **2014**, *125*, 10–21. [CrossRef]

62. Hayashi, K.I.; Fujisawa, H.; Holland, H.D.; Ohmoto, H. Geochemistry of approximately 1.9 Ga sedimentary rocks from northeastern Labrador, Canada. *Geochim. Cosmochim. Acta* **1997**, *61*, 4115–4137. [CrossRef]

63. He, B.; Xu, Y.-G.; Zhong, Y.-T.; Guan, J.-P. The Guadalupian Lopingian boundary mudstones at Chaotian (SW China) are clastic rocks rather than acidic tuffs: Implication for a temporal coincidence between the end-Guadalupian mass extinction and the Emeishan volcanism. *Lithos* **2010**, *119*, 10–19. [CrossRef]

64. Mclennan, S.M. Geochemical approaches to sedimentation, provenance, and tectonics. Processes Controlling the Composition of Clastic Sediments. *Geol. Soc. Am. Spec. Pap.* **1993**, *284*, 21.

65. Yang, N.; Tang, S.-H.; Zhang, S.-H.; Xi, Z.-D.; Li, J.; Yuan, Y.; Guo, Y.-Y. In seam variation of element-oxides and trace elements in coal from the eastern Ordos Basin, China. *Int. J. Coal Geol.* **2018**, *197*, 31–41. [CrossRef]

66. Bi, J.-H.; Ge, W.-C.; Yang, H.; Wang, Z.-H.; Dong, Y.; Liu, X.-W.; Ji, Z. Age, petrogenesis, and tectonic setting of the Permian bimodal volcanic rocks in the eastern Jiamusi Massif, NE China. *J. Asian Earth Sci.* **2017**, *134*, 160–175. [CrossRef]

67. Tang, J.; Xu, W.-L.; Wang, F.; Wang, W.; Xu, M.-J.; Zhang, Y.-H. Geochronology and geochemistry of Neoproterozoic magmatism in the Erguna Massif, NE China: Petrogenesis and implications for the breakup of the Rodinia supercontinent. *Precambrian Res.* **2013**, *224*, 597–611. [CrossRef]

68. Ji, W.; Xu, W.; Yang, D.; Pei, F.; Jin, K.; Liu, X. Chronology and geochemistry of volcanic rocks in the Cretaceous Suifenhe Formation in eastern Heilongjiang, China. *Acta Geol. Sin.* **2007**, *81*, 266–277.

69. Luan, J.-P.; Xu, W.-L.; Wang, F.; Wang, Z.-W.; Guo, P. Age and geochemistry of Neoproterozoic granitoids in the Songnen–Zhangguangcai Range Massif, NE China: Petrogenesis and tectonic implications. *J. Asian Earth Sci.* **2017**, *148*, 265–276. [CrossRef]

70. Sun, M.; Chen, H.; Zhang, F.; Wilde, S.; Minna, A.; Lin, X.; Yang, S. Cretaceous provenance change in the Hegang Basin and its connection with the Songliao Basin, NE China: Evidence for lithospheric extension driven by palaeo-Pacific roll-back. *Geol. Soc. Lond. Spec. Publ.* **2015**, *413*, 91–117. [CrossRef]

71. Zhao, L.; Dai, S.; Nechaev, V.P.; Nechaeva, E.V.; Graham, I.T.; French, D.; Sun, J. Enrichment of critical elements (Nb-Ta-Zr-Hf-REE) within coal and host rocks from the Datanhao mine, Daqingshan Coalfield, northern China. *Ore Geol. Rev.* **2019**, *111*, 102951. [CrossRef]

72. Dai, S.; Bechtel, A.; Eble, C.F.; Flores, R.M.; French, D.; Graham, I.T.; Hood, M.M.; Hower, J.C.; Korasidis, V.A.; Moore, T.A.; et al. Recognition of peat depositional environments in coal: A review. *Int. J. Coal Geol.* **2020**, *219*, 103383. [CrossRef]

73. Wang, A.; Wang, Z.; Liu, J.; Xu, N.; Li, H. The Sr/Ba ratio response to salinity in clastic sediments of the Yangtze River Delta. *Chem. Geol.* **2021**, *559*, 119923. [CrossRef]

74. Wang, A.H. Discriminant effect of sedimentary environment by the Sr/Ba ratio of different existing forms. *Acta Sedimentol. Sin.* **1996**, *14*, 168–173, (In Chinese with English abstract).

75. Ge, W.; Wu, F.; Zhou, C.; Zhang, J. Porphyry Cu-Mo deposits in the eastern Xing'an-Mongolian Orogenic Belt: Mineralization ages and their geodynamic implications. *Chin. Sci. Bull.* **2007**, *52*, 3416–3427. [CrossRef]

76. Zhou, Y.; Ren, Y.; Tang, D.; Bohor, B. Characteristics of zircons from volcanic ash-derived tonsteins in Late Permian coal fields of eastern Yunnan, China. *Int. J. Coal Geol.* **1994**, *25*, 243–264. [CrossRef]

77. Zhou, Y.; Bohor, B.F.; Ren, Y. Trace element geochemistry of altered volcanic ash layers (tonsteins) in Late Permian coal-bearing formations of eastern Yunnan and western Guizhou Provinces, China. *Int. J. Coal Geol.* **2000**, *44*, 305–324. [CrossRef]

Article

Study on the Mineralogical and Geochemical Characteristics of Arsenic in Permian Coals: Focusing on the Coalfields of Shanxi Formation in Northern China

Liqun Zhang [1,2], Liugen Zheng [1,2,]* and Meng Liu [1]

[1] School of Resources and Environmental Engineering, Anhui University, Hefei 230093, China; x21101001@stu.ahu.edu.cn (L.Z.); x19101001@stu.ahu.edu.cn (M.L.)

[2] Anhui Province Engineering Laboratory for Mine Ecological Remediation, Anhui University, Hefei 230601, China

* Correspondence: lgzheng@ustc.edu.cn; Tel./Fax: +86-551-63861471

Abstract: The Huainan Coalfield is a typical multi-coal seam coalfield. In order to systematically investigate the distribution, occurrence, and integration of arsenic (As) in Shanxi coal, 26 coal samples and three rock samples were collected in the No. 1 coal seam of Huainan coalfield. The minerals, major element oxides, and As were analyzed by X-ray diffraction (XRD), scanning electron microscopy (SEM), polarized light microscopy, X-ray fluorescence spectroscopy (XRF) and inductively coupled plasma-mass spectrometry (ICP-MS). The results indicated that the coals of Shanxi Formation were characterized by very low ash yields and low total sulfur contents. The identified minerals by XRD in the studied coals are dominated by kaolinite, quartz, calcite, and a lesser amount of pyrite. The As content ranges from 10.33 mg/kg to 95.03 mg/kg, with an average of 44.74 mg/kg. Compared with world coals, the studied coals have higher contents of As, which are characterized by enrichment. Based on statistical analyses, As shows an affinity to ash yield and possible association with silicate minerals. The contents of As in all occurrence fractions were ranked from high to low as follows: residual > Fe-Mn oxides > organic > exchangeable > carbonate. Using B, $w(Sr)/w(Ba)$ and $w(B)/w(Ga)$ geochemical parameter results to invert the depositional environment of the Huainan Shanxi Formation, a suitable coal-forming environment can cause relatively enriched As in coal.

Keywords: coalfield; arsenic; enrichment; depositional environment

Citation: Zhang, L.; Zheng, L.; Liu, M. Study on the Mineralogical and Geochemical Characteristics of Arsenic in Permian Coals: Focusing on the Coalfields of Shanxi Formation in Northern China. *Energies* **2022**, *15*, 3185. https://doi.org/10.3390/en15093185

Academic Editors: Jing Li, Yidong Cai, Lei Zhao and Rajender Gupta

Received: 11 March 2022
Accepted: 21 April 2022
Published: 27 April 2022

Publisher's Note: MDPI stays neutral with regard to jurisdictional claims in published maps and institutional affiliations.

1. Introduction

China is the world's largest energy consumer, with coal as the main fuel for its energy consumption [1]. Coal contains potentially harmful trace elements, of which As is receiving increasing attention due to its volatility, toxicity, and carcinogenicity [2,3]. As one of the volatile elements in coal, As will be released in the process of coal processing and combustion, which may seriously affect the soil and water quality around the mining area, interfere with the normal function of the immune system, and pose a threat to human health. Therefore, As concentrations and its mode of occurrence in coal have been studied by several researchers during the last three decades [4–7]. These studies show that As enrichments in coal could be controlled by several parameters, such as presence of As-bearing sulfide minerals, clastic influx and/or influence of seawater into paleomires, redox conditions within paleomires, or influence of hydrothermal solutions during coalification.

The previous studies show that Chinese coals display variable ash yields, and some studies reported up to 10 mg/kg As concentrations; however, the As content is significantly different in different coal ages, regions and coal types [8]. In China, high-As coal is widely distributed in point-like distribution, mainly located in the three northeastern provinces of Henan Yima, Shanxi Datong, Guangxi Nanning, Gansu and parts of Yunnan [9]. In addition, the occurrence modes and enrichment origins of As in different regions and different coal types in China are different. Some scholars have found that the most important geological sources of As in the Santanghu coalfield are related to penetration of fissure-hydrothermal solutions and groundwater into coal seams [10]. It is found that the accumulation of As in the peat mire environment of the Guizhou No. 6, 7, 23, and 27 coal seams is mainly controlled by the marine influence during and/or after peat accumulation [11]. China is rich in coal reserves, and the regions involved in many studies are relatively scattered. Specific to the Huainan Coalfield, previous studies have mainly focused on the Permian Upper Shihezi and Lower Shihezi formations [12–15]. However, with the increase in demand for coal resources, the mining of the Huainan Coalfield has gradually shifted from shallow coal to the deep Shanxi Formation. The influence of As on the environment, and the difficulty and approach to utilize or remove specific trace elements, are mainly dependent on the occurrence of elements [16]. Different occurrence states have a great impact on the migration, transformation, and bioavailability of As in the natural environment. Therefore, it is necessary for us to conduct a systematic study on the As in Shanxi Formation coal.

Some studies have inferred the depositional environment of the Permian coal-bearing strata in the Huainan Coalfield by using the characteristics of geochemistry, mineralogy, paleontology, sedimentary structure, lithology, and coalbeds [17–19]. The Shanxi Formation was an important peat accumulation period in the Huainan Coalfield, and it is distributed in the foreland fold-thrust belt and its frontal area of the Dabie-Sulu orogenic belt. In the early stages of coalification and post-generation rock formation, it experienced multiple stages of strong regional tectonic movements and more frequent seaward-regressive events. A systematic investigation was conducted from the No. 1 seam in the Huainan Coalfield to provide basic data on the characteristics of the coal quality and the geochemical composition. The purpose of this study is to: (1) investigate the chemical characteristics and mineral distribution of Shanxi coal, (2) analyze the occurrence characteristics of As in the coal seam and explore its depositional environment, and (3) discuss the origin of As enrichment. Collectively, the results of this study could provide a theoretical basis for the processing and utilization of associated resources and potential evaluation in the Huainan coalfield.

2. Geological Background

The Huainan coalfield is an important coal production area in east China. It is located in the north-central part of Anhui Province. It extends into the Chuxian area in the east and extends to the vicinity of Fuyang in the west. The coal mining area is 180 km in length, 15–20 km in width and covers an area of 3200 km^2 (Figure 1). The coal field is a complex syncline structure. The structural features of the main body of the complex syncline in Huainan are distributed in the east-west direction due to the squeezing action of the compressive stress in the north and south. A series of compression-torsional inverse faults, thrust faults and large nappe bodies are developed on the north and south wings of the complex syncline, and the imbricate structure of the two wings of the complex syncline is formed, which makes some strata in the south wing reverse upright.

The Carboniferous-Permian period was an important peat-forming period in the study area [20]. As one of the five major coal fields in China, the Huainan Coalfield is a typical Permian multi-peat forming environment; in turn, up to 21 coal seams are located in the Permian sequences (Figure 2). The coal-bearing strata in the Huainan Coalfield include the Benxi Formation of the Late Carboniferous, the Taiyuan Formation, the Shanxi Formation and the Lower Shihezi Formation of the Early Permian, and the Upper Shihezi Formation of the Late Permian. The Upper Shihezi Formation, the Lower Shihezi Formation, and the Shanxi Formation constitute the main mineable coal-bearing sequences in Huainan, which are a complete deltaic system developed on the offshore bay [2]. Among them, the Shanxi Formation was integrated into the Taiyuan Formation and was in contact with the Lower Shihezi Formation, and the mineable coal seams in this formation are the No. 1 and No. 3 coal seams. The Shanxi Formation shows a set of detrital coal-bearing sequences dominated by deltaic sediments, with a complete cyclonic structure of pre-triangle, delta front, and delta plain deposits. The lower part is a prodelta facies deposit, and its lithology is mainly siltstone-silty mudstone and dark mudstone. The No. 1 and No. 3 coal seams in the main mineable coal seams are developed in this sedimentary system.

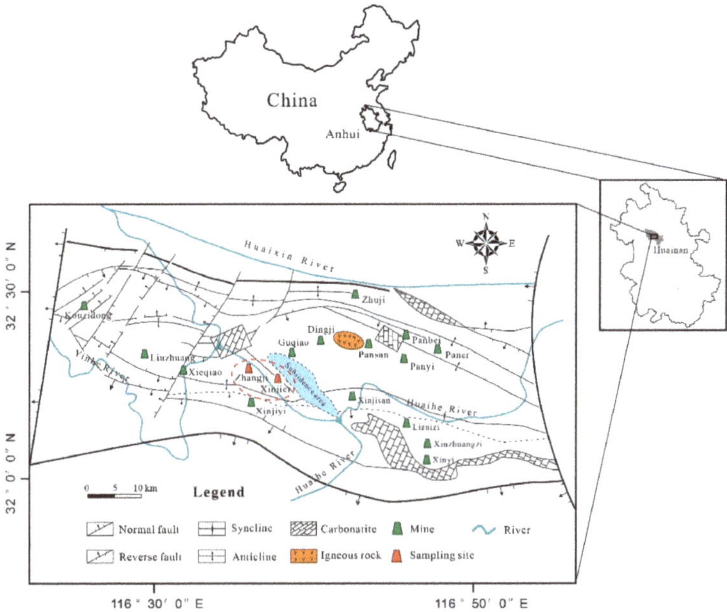

Figure 1. Location map of Huainan Coalfield, China.

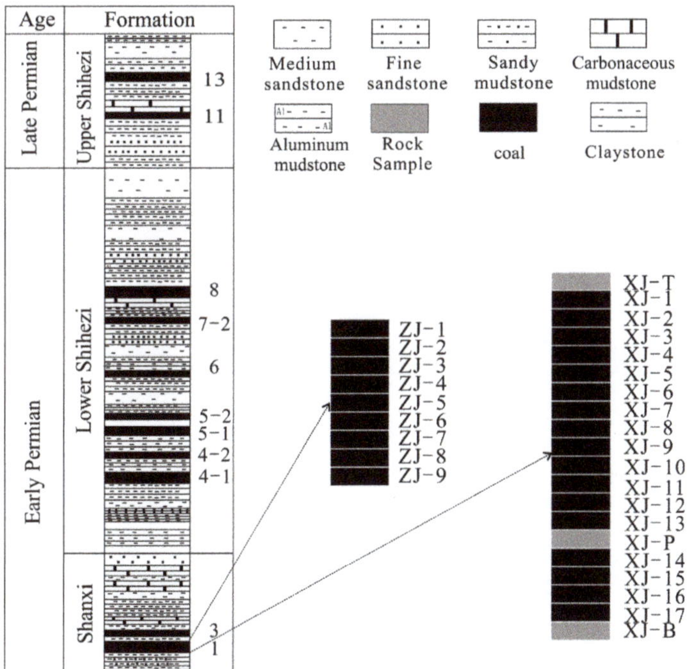

Figure 2. Generalized stratigraphic column and specific sampling points in the study area.

3. Methodology

3.1. Sampling

During the coal exploration studies of the Shanxi Formation coals, a total of twenty-nine samples (including coal, roof, floor, and parting samples) were collected from the Zhangji mine and Xinjier mine by using channel sampling, with a sample taken every 10 cm interval. Among them, the numbers of coal samples are ZJ-1–ZJ-9 and XJ-1–XJ-17, and the roof, the parting, and the floor samples are numbered XJ-T, XJ-P, and XJ-B respectively (Figure 2). All samples were immediately stored in polyethylene bags to prevent contamination, oxidation, and loss of moisture. They were brought back to the laboratory and let dry naturally, then pass through a 200-mesh sieve after grinding.

3.2. Analysis

According to the Chinese National Standard GB/T211-2008, the moisture (M), ash yield (A_d) and volatile matter (V) in coal were measured by automatic industrial analyzer (SDTGA5000a, Sundy, Changsha, China). The sulfate sulfur (S_s), pyritic sulfur (S_p), and organic sulfur (S_o) were determined following GB/T215-2003, and the total sulfur (S_t) was determined following GB/T 214-2007.

Phase-mineral composition of coal was determined by XRD (SmartLab 9, Rigaku Industrial Corporation, Osaka, Japan), acceleration voltage $\leq$ 45 kV, tube flow $\leq$ 200 mA, power $\leq$ 9 kW, scanning range was $2°{\sim}160°$ (2θ), 2θ angle indication error was $0.017°$, resolution was 27%, and diffraction intensity stability was 1.1%. The fine structure observation was analyzed by SEM (S-4800, Hitachi Corporation, Tokyo, Japan); the secondary electron resolution was 1.0 nm (15 kV), magnification was $20{\sim}800$, acceleration voltage was $0.1{\sim}30$ kV, and beam was 1 pA$\sim$2 nA. The microscopic morphology of minerals was observed by polarized light microscope (BX53, Olympus, Tokyo, Japan); the microscope condition was manual focusing, the lifting range was 50 mm, and the visual magnification was $40\times{-}500\times$.

The major oxides were determined by X-ray fluorescence spectrometry (ZSX Primus II type, Rigaku Industrial Corporation, Tokyo, Japan). The 0.1 g sample was accurately weighed in an acidic mixture and digested into a transparent solution on a hot plate at 110 °C. Then, each solution was filtered through a 0.45 μm membrane and made up to 25 mL with deionized water with 5% HNO_3. The trace elements (As, B, Sr, Ba, Ga) in the coal were determined by ICP-MS (Agilent 7500cx, Agilent, Palo Alto, CA, USA). The working parameters of the suppressor were: RF power 1500 W, auxiliary gas (Ar) flow 0.90 L/min, atomizer (Ar) flow 0.25 L/min, and the error analysis was -1.775 ± 2.745. The interference of ArCl to element As is eliminated by collision cell, and the flow rate of collision gas (He) is 0.7 mL/min. The chemical forms of As were analyzed by a sequential chemical extraction procedure (Table 1), and the recovery was $97.2{\sim}101.7\%$. The accuracy of As was determined by standards reference material GBW11116.

Table 1. Sequential chemical extraction procedure used for arsenic speciation. Adapted with permission from Elsevier, 2022 [2].

Step	Speciation	Extractant	Extraction Conditions	Cellulose Filter
F1	exchangeable	1.00 g sample + 8 mL NaOAc (1 M, pH = 8.2)	oscillate at room temperature of (25 ± 2) °C for 1 h, centrifuge	0.1 μm
F2	carbonate	sample recovered in F1 + HOAc (1 M, pH = 5.0)	stir until the reaction is complete at room temperature, then centrifuge	0.1 μm
F3	Fe-Mn oxides	sample recovered in F2 + 20 mL of 0.3 M $Na_2S_2O_4$ + 0.175 M Na-citrate + 0.025H-citrate	occasional stirring at 96 ± 3 °C, then centrifuge	0.1 μm
F4	organic	sample recovered in F3 + ① 3 mL of 0.02 M HNO_3 + 5 mL of H_2O_2 (pH = 2); ② 3 mL of 30% H_2O_2 (pH = 2, with HNO_3); ③ 5 mL of 3.2 M NH_4OAc	① 2 h at 85 ± 2 °C ② 3 h at 85 ± 2 °C ③ 0.5 h continuous stirring, then centrifuge	0.1 μm
F5	residual	tailings recovered in F4 + 5 mL HNO_3 + 5 mL HF	digestion at 110 °C to clear liquid, and then the cover was lifted at 90 °C to remove the acid	0.1 μm

4. Results

4.1. Standard Coal Characteristics

The moisture (M) content of coal in the Zhangji Mine is 1.79~2.17%, with an average of 1.98%. The ash yield (A_d) of coals in the Zhangji Mine is 5.94~13.30%, with an average of 8.38%. The content of volatile matter (V) in Zhangji Mine coal is 29.97~38.04%, with an average of 33.72%. The total sulfur (S_t) content is 0.11~0.37%, with an average of 0.16%, and the pyritic sulfur (S_p), sulfate sulfur (S_s), and organic sulfur (S_o) accounts for 64%, 12%, and 24% of total sulfur, respectively (Table 2).

Table 2. Main coal quality parameter values of coal in the Shanxi Formation in the Huainan Coalfield (%).

Samples	M (%)	A_d (%)	V (%)	S_t (%)	S_p (%)	S_s (%)	S_o (%)
ZJ-1	2.12	5.65	38.04	0.37	0.26	0.02	0.09
ZJ-2	1.79	13.3	29.97	0.20	0.14	0.02	0.04
ZJ-3	2.17	7.44	35.61	0.25	0.17	0.03	0.05
ZJ-4	2.02	6.02	36.43	0.30	0.20	0.05	0.05
ZJ-5	1.81	5.94	33.26	0.22	0.15	0.04	0.03
ZJ-6	2.07	6.03	36.26	0.28	0.18	0.04	0.06
ZJ-7	1.93	11.90	31.60	0.17	0.08	0.02	0.07
ZJ-8	1.97	9.70	31.03	0.25	0.15	0.03	0.07
ZJ-9	1.92	9.40	31.32	0.18	0.12	0.01	0.05
Ave (ZJ)	1.98	8.38	33.72	0.25	0.16	0.03	0.06
XJ-1	2.05	9.73	33.43	1.70	1.21	0.05	0.44
XJ-2	1.76	5.25	34.61	1.05	0.67	0.03	0.35
XJ-3	1.94	12.28	31.8	2.45	0.64	0.09	1.72
XJ-4	1.64	14.34	27.38	2.40	0.77	0.13	1.50
XJ-5	1.97	11.67	29.58	0.37	0.19	0.02	0.16
XJ-6	1.63	7.55	28.16	0.41	0.27	0.05	0.09
XJ-7	1.57	6.66	30.77	0.33	0.22	0.03	0.08
XJ-8	1.86	8.08	32.02	0.37	0.24	0.04	0.09
XJ-9	1.36	14.69	25.66	0.31	0.21	0.03	0.07
XJ-10	2.04	9.01	24.89	0.34	0.28	0.04	0.02
XJ-11	1.90	13.96	30.02	0.30	0.23	0.03	0.04
XJ-12	1.80	7.70	30.93	0.39	0.29	0.06	0.04
XJ-13	1.99	6.26	30.30	0.42	0.31	0.07	0.04
XJ-14	1.63	7.83	28.52	0.29	0.18	0.02	0.09
XJ-15	1.77	4.83	30.17	0.20	0.13	0.01	0.07
XJ-16	2.16	7.91	26.56	0.18	0.06	0.01	0.11
XJ-17	1.74	7.04	28.5	0.22	0.14	0.01	0.07
Ave(XJ)	1.81	9.11	29.61	0.69	0.36	0.04	0.29

M, moisture; A_d, ash yield; V, volatile matter S_t, total sulfur; S_p, pyritic sulfur; S_s, sulfate sulfur; S_o, organic sulfur.

The moisture (M) content of coal in Xinjier Mine is 1.26~2.16%, with an average of 1.81%. The ash yields (A_d) of coals in Xinjier Mine is 4.83~14.69%, with an average of 9.11%. The content of volatile matter (V) in Xinjier Mine coal is 24.89~34.61%, with an average of 29.61%. The total sulfur (S_t) content is 0.18~2.45%, with an average of 0.69%, and the pyritic sulfur (S_p), sulfate sulfur (S_s), and organic sulfur (S_o) accounts for 52.17%, 5.80%, and 42.03% of total sulfur, respectively. Among them, the total sulfur content in the XJ-1–XJ-4 areas are higher (Table 2). The coal samples from both mines could be classified as ultra-low moisture, low ash yield, medium-high volatile, and low-sulfur according to Chinese National Standards MT/850-2000 and GB/T 1522.4-2010.

4.2. Mineralogical Compositions

According to the XRD analyses results (Figure 3), the identified minerals in the Shanxi Formation raw coal samples are mainly kaolinite ($Al_4[Si_4O_{10}](OH)_8$), calcite ($CaCO_3$), quartz (SiO_2), and a small amount of pyrite (FeS_2). The microscopic morphology of pyrite is the aggregation of spherical, nodular, granular, framboidal, and fine-grained pyrite (Figure 4b,c). Banded (Figure 4d) and cell-filled kaolinite (Figure 4e) closely co-existed with pyrite (Figure 4f), formed at the same time as pyrite, and belong to the syngenetic minerals in the early diagenetic stage. Calcite is distributed in veins (Figure 4g) or filled with organic cell cavities (Figure 4h), which indicates epigenetic origin. Quartz is identified as filling the cell cavities of kinetoplastids (Figure 4i), indicating precipitation during peatification or early diagenetic stages.

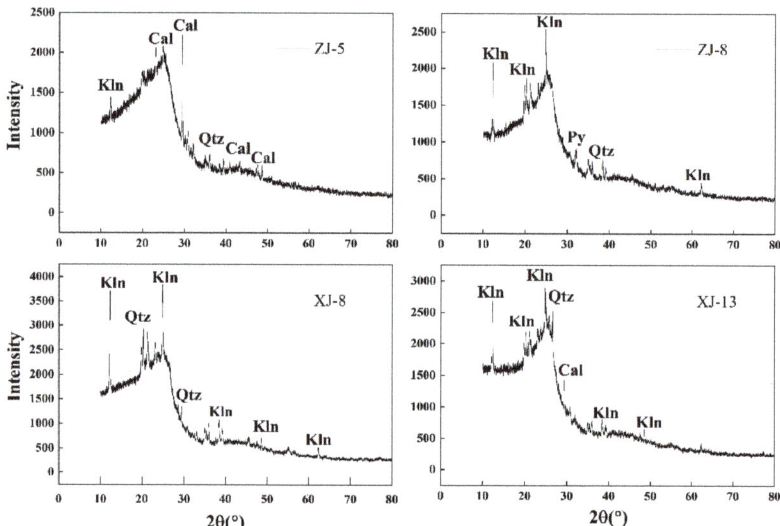

Figure 3. XRD analysis of the coal in the Shanxi Formation of Huainan (Kln—kaolinite; Qtz—quartz; Cal—calcite; Py—pyrite).

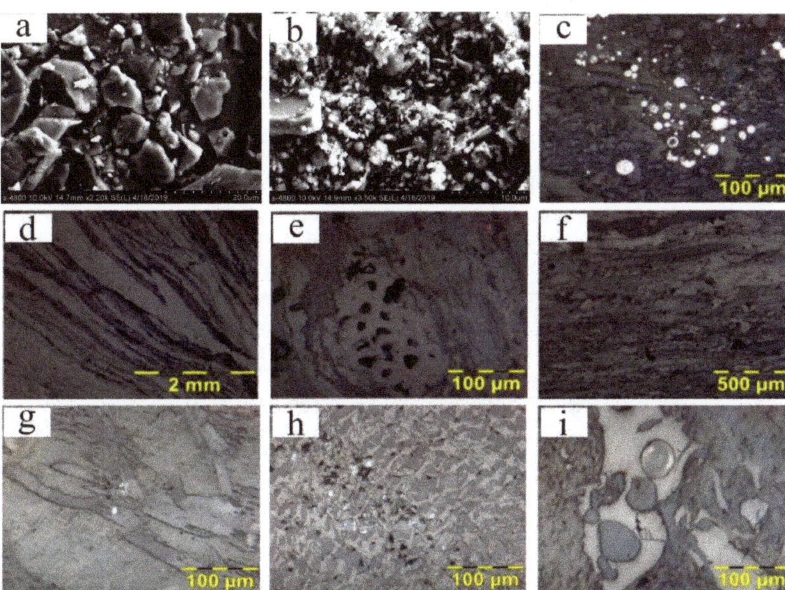

Figure 4. SEM (**a**,**b**) and optical images (**c**–**i**) of coal in Shanxi Formation. (Kln—kaolinite; Qtz—quartz; Py—pyrite) ((**a**,**b**): SEM image; (**c**): spherulitic pyrite; (**d**): banded kaolinite; (**e**): kaolinite filling the cell cavity; (**f**): co-existed with pyrite; (**g**): veined calcite; (**h**): calcite filling the cell cavity; (**i**): quartz filling the cell cavity).

4.3. Major Oxides

The average of major oxides of coal in the Zhangji mine is $Al_2O_3 > SiO_2 > Fe_2O_3 > CaO > MgO > TiO_2 > Na_2O > P_2O_5$, and the content ranges are Al_2O_3 (%) 1.81~6.21 (3.65), SiO_2 (%) 0.31~5.5 (2.89), Fe_2O_3 (%) 1.01~3.22 (1.68), CaO (%) 0.01~0.88 (0.35), MgO (%) 0.03~0.86 (0.24), TiO_2 (%) 0.05~0.51 (0.22), Na_2O (%) 0.05~0.06 (0.05), P_2O_5 (%) 0.01~0.08 (0.03). The $[w(CaO) + w(MgO) + w(Fe_2O_3)]/[w(SiO_2) + w(Al_2O_3)]$ ratio (C) of coal in the Zhangji mine ranges from 0.14~0.87 (0.41). The average of major oxides of coal in the Xinjier mine is $SiO_2 > Al_2O_3 > CaO > Fe_2O_3 > MgO > TiO_2 > P_2O_5 > Na_2O$, and the content ranges are SiO_2 (%) 2.13~24.13 (2.89), Al_2O_3 (%) 1.48~10.65 (3.65), CaO (%) 0.58~4.98 (0.35), Fe_2O_3 (%) 0.4~1.71 (1.68), MgO (%) 0.03~1.26 (0.24), TiO_2 (%) 0.11~0.48 (0.22), P_2O_5 (%) 0.02~0.17 (0.03), Na_2O (%) 0.01~0.11 (0.05). The C of coal in the Xinjier mine ranges from 0.19~0.41 (0.28) (Table 3). The main element oxides in Shanxi Formation coal are SiO_2 and Al_2O_3, and the ash yield belongs to SiO_2-Al_2O_3-Fe_2O_3-CaO.

Table 3. Content range of major oxides in the coal of Shanxi Formation in Huainan (%).

Sample	Project	Al_2O_3	SiO_2	CaO	Fe_2O_3	MgO	P_2O_5	Na_2O	TiO_2	C	Al_2O_3/TiO_2
Zhangji Mine	Min	1.81	0.31	0.01	1.01	0.03	0.01	0.05	0.05	0.14	7.89
	Max	6.21	5.5	0.88	3.22	0.86	0.08	0.06	0.51	0.87	72.25
	Ave	3.65	2.89	0.35	1.68	0.24	0.03	0.05	0.22	0.41	27.67
Xinjier Mine	Min	1.48	2.13	0.58	0.4	0.03	0.02	0.01	0.11	0.19	13.45
	Max	10.65	24.13	4.98	1.71	1.26	0.17	0.11	0.48	0.41	35.63
	Ave	4.22	6.19	1.41	0.99	0.24	0.06	0.04	0.21	0.28	20.71

$C = [w(CaO) + w(MgO) + w(Fe_2O_3)]/[w(SiO_2) + w(Al_2O_3)]$.

4.4. Content and Vertical Distribution of As

The As content in the Zhangji mine ranges from 12.51~95.03 mg/kg, with an average of 46.64 mg/kg, and the enrichment coefficient of As in Zhangji coal [21,22] (CC = content of trace elements/world average of elements in coal) is 5.62. The As content in the Xinjier mine ranges from 10.33~76.10 mg/kg, with an average of 43.73 mg/kg (Table 4), and the CC of As in Xinjier coal is 5.27. According to the Chinese coal industry standard (MT/T803-1999), the Shanxi Formation coal of Huainan Coalfield belongs to high As coal. In order to better understand the enrichment of As in coal, the As content in the coal of the Zhangji mine and Xinjier mine was compared with the Upper Shihezi and Lower Shihezi formation. The As of the Upper Shihezi formation is 6.27 mg/kg, while As in the Lower Shihezi formation is 4.81 mg/kg (Table 4). It can be seen that As shows obvious changes in the three mines. The As content in Shanxi formation coal was significantly higher than that in the Upper Shihezi and Lower Shihezi formations. However, there was a small difference in As content between the Upper Shihezi and the Lower Shihezi formations.

Table 4. Content of arsenic in coal from the Shanxi Formation in Huainan (mg/kg).

Sample	As	Sample	As
ZJ-1	42.73	XJ-1	29.14
ZJ-2	95.03	XJ-2	55.84
ZJ-3	23.54	XJ-3	12.08
ZJ-4	41.25	XJ-4	10.33
ZJ-5	13.38	XJ-5	69.16
ZJ-6	12.51	XJ-6	28.45
ZJ-7	73.96	XJ-7	41.01
ZJ-8	61.54	XJ-8	58.69
ZJ-9	55.78	XJ-9	63.05
Ave	46.64	XJ-10	42.52
XJ-T	244.65	XJ-11	48.95
XJ-P	107.97	XJ-12	76.10
XJ-B	124.65	XJ-13	57.92
Upper Shihezi [a]	6.27	XJ-14	11.13
Lower Shihezi [a]	4.81	XJ-15	33.76
Northern China [b]	3.92	XJ-16	67.60
World [c]	8.30	XJ-17	37.71
China [d]	3.79	Ave	43.73
USA [e]	24.00	-	-

[a] From Chen et al. [14]. [b] From Tian et al. [8]. [c] From Ketris and Yudovich [21]. [d] From Dai et al. [22]. [e] From Finkelman et al. [23].

The vertical distribution characteristics of As content in coal are shown in Figure 5. The content of As changed significantly, among which XJ-4 (10.33 mg/kg) had the lowest content and ZJ-2 (95.03 mg/kg) had the highest content. The As content in the roof (XJ-T), parting (XJ-P), and floor (XJ-B) of the Shanxi formation is relatively high, and respectively are 244.65 mg/kg, 107.97 mg/kg, and 124.65 mg/kg (Table 4).

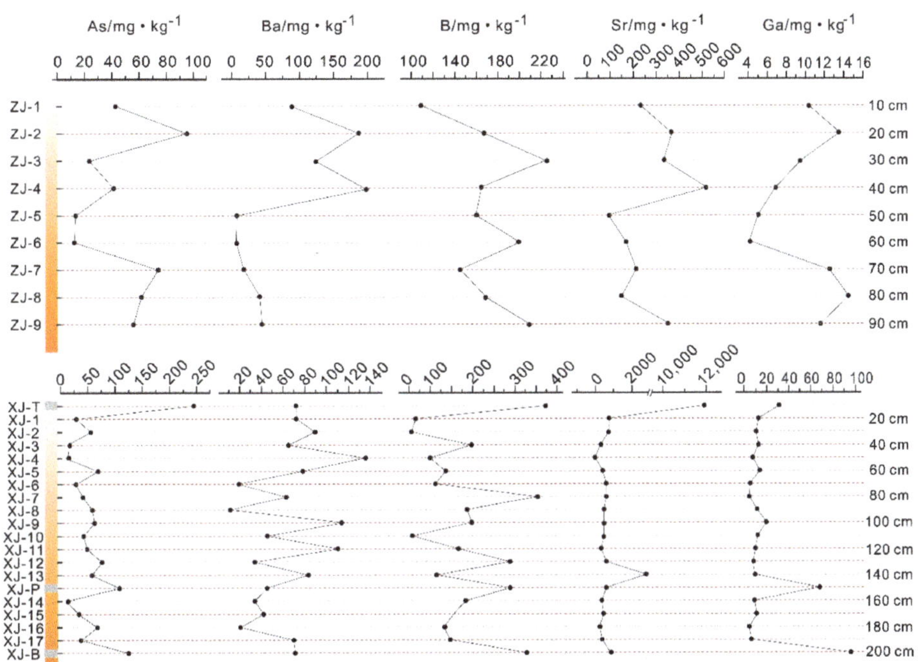

Figure 5. Vertical distribution of As, Ba, B, Sr, Ga in coal.

5. Discussion

5.1. Depositional Environment of Coal

The presence of high sulfur content is attributed to regional volcanism, peat environments, and depositional environments with strong sulfide mineralization [24,25]. Shanxi Formation coal is low-sulfur coal, but the total sulfur content in the XJ-1–XJ-4 area is relatively high (Table 2). Nevertheless, some previous studies show that the depositional environment affected by seawater may lead to the phenomenon of high sulfur in local coal seams [26]. The ash yield belongs to SiO_2-Al_2O_3-Fe_2O_3-CaO, indicating that more detritus minerals were transported to the study area and deposited on the coastal delta plain where it was open to clastic influx. The high content of SiO_2 and Al_2O_3 indicates that the minerals in the raw coal are composed of clay minerals (such as kaolinite and illite) and quartz [27]. In this study, the average ash yield of coal seams in the Shanxi Formation were considerably lower than coals from the Upper Shihezi Formation (20.12%) and Lower Shihezi Formation (21.27%) [28]. The change in ash content in the Huainan Coalfield is called "increasing strati graphically upward" [14,29].

The $[w(CaO) + w(MgO) + w(Fe_2O_3)]/[w(SiO_2) + w(Al_2O_3)]$ ratio (C) of coal can indicate the depositional environment of the peat accumulation stage. Coal with $C \leq 0.22$ could be accumulated within terrestrial environments (e.g., freshwater lake-shore), whereas $C > 0.22$ could imply transition areas between terrestrial to shallow marine (e.g., back-mangrove conditions or delta front) [30,31]. The C in the coal of the Zhangji and Xinjier Mines is 0.41 and 0.28, respectively (Table 3), indicating that the sedimentary facies present sea-land alternate facies, where marine influence into paleomires could be common. However, in practice, the conclusion of C is often affected by other sedimentary environment indicators. For instance, the presence of gastropods, ostracod fauna, and charophyta remains obtained from coal seams in northern Turkey points to the predominance of freshwater conditions. Peat is deposited in sloughs with water from karst aquifers rich in sulfate and calcium; in this case, freshwater coal exhibits characteristics similar to saltwater or ocean-influenced peat/coal [32]. Furthermore, the presence of clastic Mg-bearing silicates or clay minerals (e.g., chlorite or smectite) inputs into freshwater lakes, which increases the supply of dissolved Mg ions; in turn, the C values could be elevated [33]. Therefore, paleontological study of Shanxi Formation coal seams should be undertaken in the future for better understanding of the marine influence on peat-forming environments. Previous studies have suggested that Al_2O_3/TiO_2 is the most effective indicator for the properties of sedimentary parent rock. When the ratio of Al_2O_3/TiO_2 is 3:8, 8:21, or 21:70, it means that the sediments are formed by mafic, intermediate, or felsic igneous rocks, respectively [34,35]. The values of Al_2O_3/TiO_2 in the Huainan Shanxi Formation were widely distributed, ranging from 7.89 to 72.25, with an average of 24.19, indicating that the clastic materials in the coal mainly come from felsic rocks [36].

Even though B enrichments could be controlled by several parameters, the mass fraction of B has a good linear relationship with the paleo-salinity [23,26,37–39]. Generally, 50 mg/kg and 110 mg/kg are divided into fresh water/mildly brackish water and mildly brackish water/brackish water [40]. The highest content of B in Shanxi Formation coal is 354.60 mg/kg, with an average of 162.00 mg/kg (Table 5). Among them, the content range of B in the five coal samples of ZJ-1, XJ-1, XJ-2, XJ-4, and XJ-10 is between 50 mg/kg and 110 mg/kg, and other samples are all more than 110 mg/kg. This showed that Huainan Shanxi Formation is in the stage of mildly brackish-brackish water deposition (Figure 6a). In addition, the $w(Sr)/w(Ba)$ is a geochemical indicator that distinguishes between terrestrial and marine sedimentary environments in terrigenous clastic sediments [41]. $w(Sr)/w(Ba) < 0.6$ indicated terrestrial freshwater deposition, $w(Sr)/w(Ba)$ between 0.6~1 suggested mildly brackish water deposition, and $w(Sr)/w(Ba) > 1$ indicated brackish water deposition [41,42]. The $w(Sr)/w(Ba)$ ratio of the 26 samples in the Huainan Shanxi Formation was greater than 1 (Figure 6b), suggesting that the depositional environment of the Shanxi Formation was brackish water deposition. However, the usability of this indicator ($w(Sr)/w(Ba)$) is not widely acknowledged. Therefore, to verify this theory and accurately determine the differences between the sedimentary geochemical behaviors of Sr and Ba under different salinity conditions, it is recommended that in future studies, selective extraction of the Sr and Ba concentrations of different salinities is used to discriminate between marine and terrestrial sedimentary environments in terrigenous clastic sediments. An alternative geochemical indicator of the depositional environments is $w(B)/w(Ga)$ ratio in coals, that is, fresh water (<3), brackish water influences (3~5) and brackish water influences (>5) [43]. The values of $w(B)/w(Ga)$ in XJ-2 and XJ-10 in Shanxi Formation coal in the study area are slightly lower than the mildly brackish water/brackish water boundary value of 5, and other samples $w(B)/w(Ga)$ are greater than 5 (Figure 6c). In conclusion, the depositional environment of Shanxi Formation in Huainan is mildly brackish-brackish water deposition.

Table 5. The content range of B, Sr, Ba and Ga in coal.

Project	B (mg/kg)	Sr (mg/kg)	Ba (mg/kg)	Ga (mg/kg)	Sr/Ba	B/Ga
Min	52.32	71.29	11.74	4.22	1.65	4.81
Max	354.6	2742.71	135.69	20.37	17.72	48.44
Ave	162	585.08	72.35	9.87	5.81	20.57

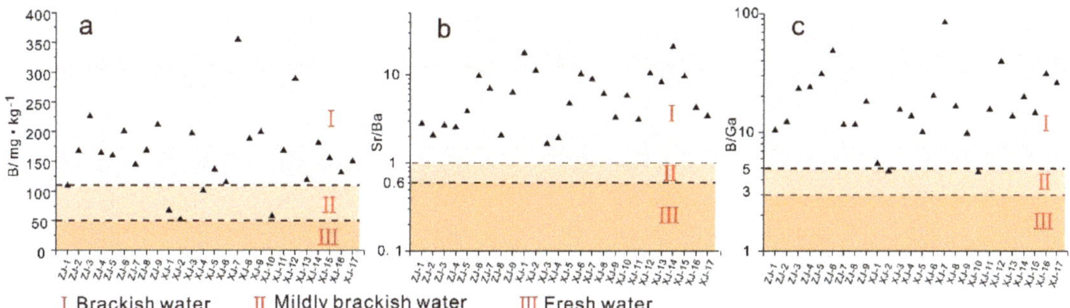

Figure 6. Judging the depositional environment of the Huainan Shanxi Formation. (**a**): B; (**b**): $w(Sr)/w(Ba)$; (**c**): $w(B)/w(Ga)$.

5.2. Geochemistry of As

The occurrence of As in coal is complicated, and it participates in the formation of inorganic or organic bound states in the structure of coal. For coal samples with a narrow ash yield range, correlation coefficients can help to infer how trace elements are present in coal in depositional environments [44]. In this study, we analyzed the correlation between As and ash, sulfur, and coal ash components, and combined with sequential chemical extraction experiment to explore the occurrence of As in the coals of Shanxi formation. Generally, As has a strong affinity for sulfur, and the relationship between As and sulfur increases and decreases simultaneously [45]. In this study, As and S contents are negatively correlated (r = −0.41), suggesting that less sulfur-arsenic minerals were present in the studied coal samples. However, the lack of As in SEM-EDX spectra does not mean that either pyrite or clay minerals do not contain As. Since the As measurement capacity is low, measurable amounts of As could not be seen in SEM-EDX spectra. Of course, the lack of As within framboidal pyrite grains are expectable, since these grains could have lower As concentrations [6,46]. On the other hand, the correlation coefficient between As and ash content was 0.53, showing a positive correlation (Figure 7a), indicating that the main carrier of As in coal might be affilated with aluminosilicate minerals in the studied samples (e.g., clay minerals). In agreement with this correlation, As is positively correlated with Al_2O_3 and TiO_2, indicating that As could be mainly affilated with clay minerals in coal samples [47,48]. According to the result of sequential chemical extraction experiment (Figure 7b), the order of As content in each speciation is residual > Fe-Mn oxides > organic > exchangeable > carbonate. The main speciation of As is residual with 81.36%. The residual is difficult to dissolve by weak acid and general solvent, and its chemical activity is weak in the environment.

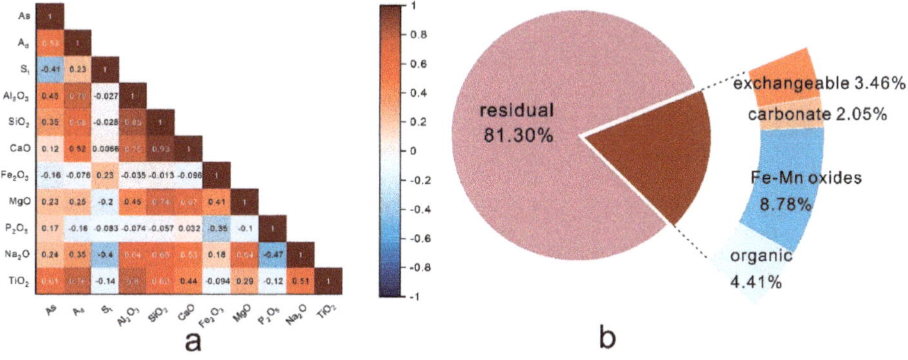

Figure 7. Correlation analysis (**a**) and speciation distribution of As (**b**) in coal.

The As distribution in the study area varies widely, which is mainly related to the peat-forming environment and mineralogical compositions of coal seams. This study shows that the Shanxi formation was deposited in a brackish environment, with decreasing marine influence along stratigraphy upward, increasing input of terrigenous detrital materials and significantly reduced content of As. Influenced by the depositional environment of Shanxi formation, there are clastic influxes of different origins in the roof (XJ-T), parting (XJ-P), and floor (XJ-B), which causes Shanxi Formation coal to have a special rock roof and good waterproof roof, parting and floor. The roof is a thick sandstone mainly composed of feldspar, and the floor is made up of aquifer-bearing Carboniferous, Ordovician, and Cambrian formations. The parting is mostly mudstone or carbonaceous mudstone, which will lead to high trace elements in the roof, parting, and floor. In the early and end stages of peat mires, the higher concentration of mineral solution seems to penetrate into paleomires, and the change of environmental conditions is not conducive to the normal growth of plants; trace elements are precipitated due to the obstructed circulation process, which leads to the higher content of trace elements in the top and bottom layers [49]. Due to the influence of depositional environment, a thin layer of mudstone and carbonaceous mudstone with high trace elements in gangue inclusion is formed.

5.3. Controlling Factors on As Enrichment

There are generally one or more particular geological factors that may influence the enrichment of trace elements in coals for different coal basins and coal forming periods [50,51]. The As content in Shanxi Formation coal in the Huainan Coalfield is obviously higher than that of the Upper Shihezi and Lower Shihezi formations, which indicates that As may have local enrichment characteristics and is closely related to the depositional environment of coal seam.

The contents of B and As in the Huainan Permian Upper Shihezi Formation (No. 11 and 13 coal seams), Lower Shihezi Formation (No. 4, 5, 6, 7, 8, and 9 coal seams), and Shanxi Formation No. 1 coal seam are shown in Figure 8 [14]. The high B contents in the No. 1 coal seam indicates that the peat marsh formed in the Shanxi Formation was greatly affected by sea water. Boron content in each coal seam of the Shihezi Formation indicates that the upper and lower Shihezi Formation are generally transitional phase or brackish water deposition environments, and the B content in No. 5 coal seam and No. 11 coal seam is lower than the fresh water/mildly brackish water boundary value 50 mg/kg, which is characterized by continental deposition.

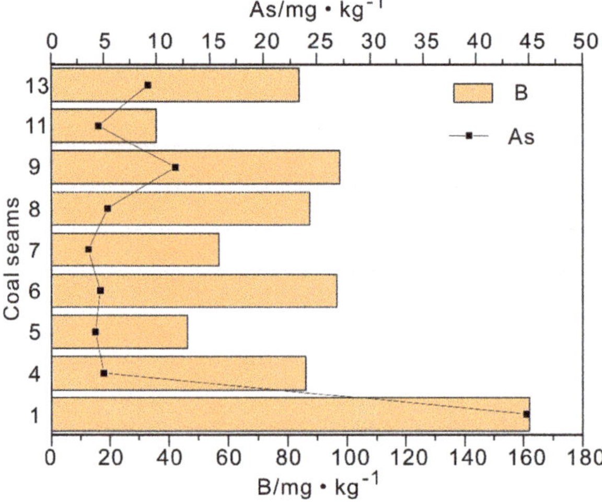

Figure 8. Content changes of B and As in different coal seams in Huainan.

During the Permian, several transgressions and regressions took place in the study area. The change trend of B contents in the longitudinal direction of the No. 1 coal seam in the Shanxi Formation to the No. 4–9 coal seams in the Lower Shihezi Formation to the No. 11 and 13 coal seams in the Upper Shihezi Formation is the same as that of As. The No. 1 coal seam of the Shanxi Formation was more affected by seawater than the coal seams of the Shihezi Formation. The S content in the coal has the overall order of Shanxi Formation > Lower Shihezi Formation > Upper Shihezi Formation. In this study, As occurs less in sulfides and more in silico-aluminate minerals. This shows that As in Shanxi Formation coal is mainly derived from terrestrial detrital sediments, which were brought into coal-forming mires by water and adsorbed into coal by humic acid, leading to enrichment of As in coal. Appropriate depositional environments can lead to relatively enriched As in coal.

6. Conclusions

The coal in the Shanxi Formation belongs to the ultra-low ash, ultra-low total moisture, medium-high volatile, low-sulfur coal category. The major minerals include quartz, kaolinite, calcite, and a small amount of pyrite. The coal in the Huainan Shanxi Fomation was mainly affected by seawater, and the detrital material in the coal mainly comes from felsic rock. The enrichment coefficient CC of the coal is 5.39, which is characterized by enrichment. Among the samples, there are clastic materials from different sources in the roof, floor, and parting, and the As content is significantly higher than that in the coal. The residual state is the main form of As, and As is mainly found in clay minerals such as aluminosilicate. In addition, the recognition results of paleo-salinity characteristics indicate that the environment as a whole is brackish-saltwater sedimentation. A suitable depositional environment can lead to the relative enrichment of As in coal. The main source of As is terrigenous detritus, which is carried into the coal-marsh by water and adsorbed into coal by humic acid, thus leading to the enrichment of As in coal.

Author Contributions: Writing—original draft, L.Z. (Liqun Zhang); Writing—review and editing, L.Z. (Liqun Zhang); Methodology, L.Z. (Liqun Zhang) and M.L.; Investigation, L.Z. (Liqun Zhang); Software, L.Z. (Liqun Zhang); Conceptualization, L.Z. (Liugen Zheng); Resources, L.Z. (Liugen Zheng); Funding acquisition, L.Z. (Liugen Zheng); Supervision, L.Z. (Liugen Zheng); Data curation, M.L. All authors have read and agreed to the published version of the manuscript.

Funding: This research was supported by the University Synergy Innovation Program of Anhui Province (GXXT-2021-017) and the National Natural Science Foundation of China (42072201). We acknowledge the editors and reviewers for polishing the language and for in-depth discussion.

Institutional Review Board Statement: Not applicable.

Informed Consent Statement: Not applicable.

Data Availability Statement: All data generated or used during the study appear in the submitted article.

Conflicts of Interest: The authors declare no conflict of interest.

References

1. Kang, Y.; Liu, G.J.; Chou, C.L.; Wong, M.H.; Zheng, L.G.; Ding, R. Arsenic in Chinese coals: Distribution, modes of occurrence, and environmental effects. *Sci. Total Environ.* **2011**, *412–413*, 1–13. [CrossRef]
2. Hu, G.Q.; Liu, G.J.; Wu, D.; Fu, B. Geochemical behavior of hazardous volatile elements in coals with different geological origin during combustion. *Fuel* **2018**, *233*, 361–376. [CrossRef]
3. Zhang, L.Q.; Zhou, H.H.; Chen, X.; Liu, G.J.; Jiang, C.L.; Zheng, L.G. Study of the micromorphology and health risks of arsenic in copper smelting slag tailings for safe resource utilization. *Ecotoxicol. Environ. Saf.* **2021**, *219*, 112321. [CrossRef] [PubMed]
4. Dai, S.F.; Zeng, R.S.; Sun, Y. Enrichment of arsenic, antimony, mercury, and thallium in a Late Permian anthracite from Xingren, Guizhou, Southwest China. *Int. J. Coal Geol.* **2006**, *66*, 217–226. [CrossRef]
5. Diehl, S.F.; Goldhaber, M.B.; Koenig, A.E.; Lowers, H.A.; Ruppert, L.F. Distribution of arsenic, selenium, and other trace elements in high pyrite Appalachian coals: Evidence for multiple episodes of pyrite formation. *Int. J. Coal Geol.* **2012**, *94*, 238–249. [CrossRef]

6. Karayiğit, A.; Bircan, C.; Mastalerz, M.; Oskay, R.; Querol, X.; Lieberman, N.; Türkmenb, I. Coal characteristics, elemental composition and modes of occurrence of some elements in the İsaalan coal (Balıkesir, NW Turkey). *Int. J. Coal Geol.* **2017**, *172*, 43–59. [CrossRef]

7. Yudovich, Y.E.; Ketris, M.P. Mercury in coal: A review: Part 1. Geochemistry. *Int. J. Coal Geol.* **2005**, *62*, 107–134. [CrossRef]

8. Tian, H.Z.; Zhu, C.Y.; Gao, J.J.; Cheng, K.; Hao, J.M.; Wang, K.; Hua, S.B.; Wang, Y.; Zhou, J.R. Quantitative assessment of atmospheric emissions of toxic heavy metals from anthropogenic sources in China: Historical trend, spatial distribution, uncertainties, and control policies. *Atmos. Chem. Phys.* **2015**, *15*, 10127–10147. [CrossRef]

9. Fu, C.; Bai, X.F.; Jiang, Y. Discussion on the relationship between the content of Arsenic and the coal quality characteristic and the Arsenic modes of occurrence in Chinese high Arsenic coal. *J. China Coal Soc.* **2012**, *37*, 96–102. (In Chinese)

10. Zhang, Y.Y.; Tian, J.J.; Feng, S.; Yang, F.; Lu, X.Y. The occurrence modes and geologic origins of arsenic in coal from Santanghu Coalfield, Xinjiang. *J. Geochem. Explor.* **2018**, *186*, 225–234. [CrossRef]

11. Li, B.; Zhuang, X.; Li, J.; Querol, X.; Font, O.; Moreno, N. Enrichment and distribution of elements in the Late Permian coals from the Zhina Coalfield, Guizhou Province, Southwest China. *J. Geochem. Explor.* **2017**, *171*, 111–129. [CrossRef]

12. Sun, R.Y.; Liu, G.J.; Zheng, L.G.; Chou, C.L. Characteristics of coal quality and their relationship with coal-forming environment: A case study from the Zhuji exploration area, Huainan coalfield, Anhui, China. *Energy* **2010**, *35*, 423–435. [CrossRef]

13. Sun, R.Y.; Liu, G.J.; Zheng, L.G.; Chou, C.L. Geochemistry of trace elements in coals from the Zhuji mine, Huainan coalfield, Anhui, China. *Int. J. Coal Geol.* **2010**, *81*, 81–96. [CrossRef]

14. Chen, J.; Liu, G.J.; Jiang, M.M.; Chou, C.L.; Li, H.; Wu, B.; Zheng, L.G.; Jiang, D.D. Geochemistry of environmentally sensitive trace elements in Permian coals from the Huainan coalfield, Anhui, China. *Int. J. Coal Geol.* **2011**, *88*, 41–54. [CrossRef]

15. Chen, S.; Wu, D.; Liu, G.; Sun, R.; Fan, X. Geochemistry of major and trace elements in Permian coal: With an emphasis on no. 8 coal seam of Zhuji coal mine, Huainan coalfield, China. *Environ. Earth Sci.* **2016**, *75*, 494. [CrossRef]

16. Dai, S.F.; Liu, J.J.; Ward, C.R.; Hower, J.C.; French, D.; Jia, S.H.; Hood, M.M.; Garrison, T.M. Mineralogical and geochemical compositions of Late Permian coals and host rocks from the Guxu Coalfield, Sichuan Province, China, with emphasis on enrichment of rare metals. *Int. J. Coal Geol.* **2016**, *166*, 71–95. [CrossRef]

17. Munir, M.A.M.; Liu, G.J.; Yousaf, B.; Ali, M.U.; Abbas, Q.; Ullah, H. Enrichment of Bi-Be-Mo-Cd-Pb-Nb-Ga, REEs and Y in the Permian coals of the Huainan Coalfield, Anhui, China. *Ore Geol. Rev.* **2018**, *95*, 431–455. [CrossRef]

18. Wang, G.; Ju, Y.; Yan, Z.; Li, Q. Pore structure characteristics of coal-bearing shale using fluid invasion methods: A case study in the Huainan–Huaibei Coalfield in China. *Mar. Pet. Geol.* **2015**, *62*, 1–13. [CrossRef]

19. Chen, S.; Wu, D.; Liu, G.; Sun, R. Raman spectral characteristics of magmatic-contact metamorphic coals from Huainan Coalfield, China. *Spectrochim. Acta Part A Mol. Biomol. Spectrosc.* **2017**, *171*, 31–39. [CrossRef]

20. Wei, Q.; Hu, B.; Li, X.; Feng, S.; Xu, H.; Zheng, K.; Liu, H. Implications of geological conditions on gas content and geochemistry of deep coalbed methane reservoirs from the Panji Deep Area in the Huainan Coalfield, China. *J. Nat. Gas Sci. Eng.* **2021**, *85*, 103712. [CrossRef]

21. Ketris, M.P.; Yudovich, Y.E. Estimations of Clarkes for Carbonaceous biolithes: World averages for trace element contents in black shales and coals. *Int. J. Coal Geol.* **2009**, *78*, 135–148. [CrossRef]

22. Dai, S.F.; Luo, Y.B.; Seredin, V.V.; Ward, C.R.; Hower, J.C.; Zhao, L.; Liu, S.D.; Zhao, C.L.; Tian, H.M.; Zou, J.H. Revisiting the late Permian coal from the Huayingshan, Sichuan, southwestern China: Enrichment and occurrence modes of minerals and trace elements. *Int. J. Coal Geol.* **2014**, *122*, 110–128. [CrossRef]

23. Finkelman, R.; Dai, S.; French, D. The importance of minerals in coal as the hosts of chemical elements: A review. *Int. J. Coal Geol.* **2019**, *212*, 103251. [CrossRef]

24. Chou, C.L. Sulfur in coals: A review of geochemistry and origins. *Int. J. Coal Geol.* **2012**, *100*, 1–13. [CrossRef]

25. Dai, S.F.; Bechtel, A.; Eble, C.; Flores, R.; French, D.; Graham, I.T.; Hood, M.M.; Hower, J.C. Recognition of peat depositional environments in coal: A review. *Int. J. Coal Geol.* **2020**, *219*, 103383. [CrossRef]

26. Karayigit, A.I.; Atalay, M.; Oskay, R.G.; Córdoba, P.; Querol, X.; Bulut, Y. Variations in elemental and mineralogical compositions of Late Oligocene, Early and Middle Miocene coal seams in the Kale-Tavas Molasse sub-basin, SW Turkey. *Int. J. Coal Geol.* **2020**, *218*, 103366. [CrossRef]

27. Munir, M.A.M.; Liu, G.J.; Yousaf, B.; Ali, M.U.; Abbas, Q. Enrichment and distribution of trace elements in Padhrar, Thar and Kotli coals from Pakistan: Comparison to coals from China with an emphasis on the elements distribution. *J. Geochem. Explor.* **2018**, *185*, 153–169. [CrossRef]

28. Ding, D.S.; Liu, G.J.; Fu, B.; Qi, C.C. Characteristics of the coal quality and elemental geochemistry in Permian coals from the Xinjier mine in the Huainan Coalfield, north China: Influence of terrigenous inputs. *J. Geochem. Explor.* **2018**, *186*, 50–60.

29. Fu, B.; Liu, G.J.; Liu, Y.; Cheng, S.W.; Qi, C.C.; Sun, R.Y. Coal quality characterization and its relationship with geological process of the Early Permian Huainan coal deposits, southern North China. *J. Geochem. Explor.* **2016**, *166*, 33–44. [CrossRef]

30. Jiang, Y.F.; Zhao, L.; Zhou, G.Q.; Wang, X.B.; Zhao, L.X.; Wei, J.P.; Song, H.J. Petrological, mineralogical, and geochemical compositions of Early Jurassic coals in the Yining Coalfield, Xinjiang, China. *Int. J. Coal Geol.* **2015**, *152*, 47–67. [CrossRef]

31. Li, C.H.; Liang, H.D.; Wang, S.K.; Liu, G.J. Study of harmful trace elements and rare earth elements in the Permian tectonically deformed coals from Lugou Mine, North China Coal Basin, China. *J. Geochem. Explor.* **2018**, *190*, 10–25. [CrossRef]

32. Oskay, R.G.; Christanis, K.; Inaner, H.; Salman, M.; Taka, M. Palaeoenvironmental reconstruction of the eastern part of the Karapınar-Ayrancı coal deposit (Central Turkey). *J. Geochem. Explor.* **2016**, *163*, 100–111. [CrossRef]

33. Dill, H.G. A geological and mineralogical review of clay mineral deposits and phyllosilicate ore guides in Central Europe—A function of geodynamics and climate change. *Ore Geol. Rev.* **2020**, *119*, 103304. [CrossRef]

34. Dai, S.F.; Hower, J.C.; Ward, C.R.; Guo, W.M.; Song, H.J.; O'Keefe, J.M.K.; Xie, P.P.; Hood, M.M.; Yan, X.Y. Elements and phosphorus minerals in the middle Jurassic inertinite-rich coals of the Muli Coalfield on the Tibetan Plateau. *Int. J. Coal Geol.* **2015**, *144–145*, 23–47. [CrossRef]

35. He, B.; Xu, Y.G.; Zhong, Y.T.; Guan, J.P. The Guadalupian-Lopingian boundary mudstones at Chaotian (SW China) are clastic rocks rather than acidic tuffs: Implication for a temporal coincidence between the end-Guadalupian mass extinction and the Emeishan volcanism. *Lithos* **2010**, *119*, 10–19. [CrossRef]

36. Song, D.Y.; Li, C.H.; Song, B.Y.; Yang, C.B.; Li, Y.B. Geochemistry of mercury in the Permian tectonically deformed coals from Peigou mine, Xinmi coalfield, China. *Acta Geol. Sin.-Engl. Ed.* **2017**, *91*, 2243–2254. [CrossRef]

37. Beaton, A.P.; Goodarzi, F.; Potter, J. The petrography, mineralogy and geochemistry of a paleocene lignite from southern Saskatchewan, Canada. *Int. J. Coal Geol.* **1991**, *17*, 117–148. [CrossRef]

38. Chen, J.; Chen, P.; Yao, D.; Liu, Z.; Wu, Y.; Liu, W.; Hu, Y. Mineralogy and geochemistry of Late Permian coals from the Donglin Coal Mine in the Nantong coalfield in Chongqing, southwestern China. *Int. J. Coal Geol.* **2015**, *149*, 24–40. [CrossRef]

39. Dai, S.F.; Liu, J.; Ward, C.R.; Hower, J.C.; Xie, P.; Jiang, Y. Petrological, geochemical, and mineralogical compositions of the low-Ge coals from the Shengli Coalfield, China: A comparative study with Ge-rich coals and a formation model for coal-hosted Ge ore deposit. *Ore Geol. Rev.* **2015**, *71*, 318–349. [CrossRef]

40. Alastuey, A.; Jiménez, A.; Plana, F.; Querol, X.; Suárez-Ruiz, I. Geochemistry, mineralogy, and technological properties of the main Stephanian (Carboniferous) coal seams from the Puertollano Basin, Spain. *Int. J. Coal Geol.* **2001**, *45*, 247–265. [CrossRef]

41. Wang, A.; Wang, Z.; Liu, J.; Xu, N.; Li, H. The Sr/Ba ratio response to salinity in clastic sediments of the Yangtze River Delta. *Chem. Geol.* **2021**, *559*, 119923. [CrossRef]

42. Dai, S.; Hou, X.; Ren, D.; Tang, Y. Surface analysis of pyrite in the no. 9 coal seam, Wuda coalfield, Inner Mongolia, China, using high-resolution time-of-flight secondary ion mass-spectrometry. *Int. J. Coal Geol.* **2003**, *55*, 139–150. [CrossRef]

43. Hu, X.F.; Liu, Z.J.; Liu, R.; Liu, D.Q.; Zhang, M.M.; Xu, S.C.; Meng, Q.T. Trace Element Characteristics of Eocene Jijuntun Formation and the Favorable Metallogenic Conditions of Oil Shale in Fushun Basin. *J. Jilin Univ.* **2012**, *42*, 60–71.

44. Eskenazy, G.; Finkelman, R.B.; Chattarjee, S. Some considerations concerning the use of correlation coefficients and cluster analysis in interpreting coal geochemistry data. *Int. J. Coal Geol.* **2010**, *83*, 491–493. [CrossRef]

45. Xie, H.; Nie, A.G. The modes of occurrence and washing floatation characteristic of arsenic in coal from Western Guizhou. *J. China Coal Soc.* **2010**, *35*, 117–121.

46. Hower, J.C.; Campbell, J.L.; Teesdale, W.J.; Nejedly, Z.; Robertson, J.D. Scanning proton microprobe analysis of mercury and other trace elements in Fe-sulfides from a Kentucky coal. *Int. J. Coal Geol.* **2008**, *75*, 88–92. [CrossRef]

47. Sun, Y.Z.; Zhao, C.L.; Li, Y.H.; Wang, J.X.; Liu, S.M. Li distribution and mode of occurrences in Li-bearing coal seam # 6 from the Guanbanwusu Mine, Inner Mongolia, Northern China. *Energy Explor. Exploit.* **2012**, *30*, 109–130.

48. Wang, P.P.; Ji, D.P.; Yang, Y.C.; Zhao, L.X. Mineralogical compositions of Late Permian coals from the Yueliangtian mine, western Guizhou, China: Comparison to coals from eastern Yunnan, with an emphasis on the origin of the minerals. *Fuel* **2016**, *181*, 859–869. [CrossRef]

49. Liu, G.J.; Yang, P.Y.; Peng, Z.C.; Wang, G.L. Geochemistry of trace elements from the No.3 coal seam of Shanxi Formation in the Yanzhou mining district. *Geochimica* **2003**, *32*, 255–262.

50. Jiu, B.; Huang, W.; Mu, N. Mineralogy and elemental geochemistry of Permo-Carboniferous Li-enriched coal in the southern Ordos Basin, China: Implications for modes of occurrence, controlling factors and sources of Li in coal. *Ore Geol. Rev.* **2022**, *141*, 104686. [CrossRef]

51. Oboirien, B.O.; Thulari, V.; North, B.C. Enrichment of trace elements in bottom ash from coal oxy-combustion: Effect of coal types. *Appl. Energy* **2016**, *177*, 81–86. [CrossRef]

Article

The Influence of Fracturing Fluid Volume on the Productivity of Coalbed Methane Wells in the Southern Qinshui Basin

Wenwen Chen [1], Xiaoming Wang [1,*], Mingkai Tu [1], Fengjiao Qu [2], Weiwei Chao [3], Wei Chen [4] and Shihui Hou [5]

1   Key Laboratory of Tectonics and Petroleum Resources, China University of Geosciences, Wuhan 430074, China
2   Engineering Technology Research Institute, PetroChina Huabei Oilfield Company, Renqiu 062550, China
3   China United Coalbed Methane Company Limited, Taiyuan 030000, China
4   CBM Exploration and Development Branch, PetroChina Huabei Oilfield Company, Jincheng 047000, China
5   Laboratory of Geotechnical Engineering, Jinggangshan University, Ji'an 343000, China
*   Correspondence: sunwxm@cug.edu.cn

Citation: Chen, W.; Wang, X.; Tu, M.; Qu, F.; Chao, W.; Chen, W.; Hou, S. The Influence of Fracturing Fluid Volume on the Productivity of Coalbed Methane Wells in the Southern Qinshui Basin. *Energies* **2022**, *15*, 7673. https://doi.org/10.3390/en15207673

Academic Editor: Mohammad Sarmadivaleh

Received: 22 September 2022
Accepted: 14 October 2022
Published: 18 October 2022

Abstract: Hydraulic fracturing is the main technical means for the reservoir stimulation of coalbed methane (CBM) vertical wells. The design of fracturing fluid volume (FFV) is mainly through numerical simulation, and the numerical simulation method does not fully consider the water block damage caused by the leakage of fracturing fluid into the reservoir. In this work, the variance analysis method was used to analyze the production data of 1238 CBM vertical wells in the Fanzhuang block and Zhengzhuang block of the Qinshui Basin, to clarify the relationship between the FFV and the peak gas production (PGP) under the different ratios of critical desorption pressure to reservoir pressure ($R_{c/r}$), and to reveal the controlling mechanism of fracturing fluid on CBM migration. The results show that both the FFV and $R_{c/r}$ have a significant impact on gas production. When $R_{c/r} < 0.5$, the PGP decreases with the increase of the FFV, and the FFV that is beneficial to gas production is 200–500 m^3. When $R_{c/r} > 0.5$, the PGP increases first and then decreases with the increase of FFV. Specifically, the FFV that is favorable for gas production is 500–700 m^3. Excessive FFV does not significantly increase the length of fractures due to leaks in the coal reservoir. Instead, it is more likely to invade and stay in smaller pores, causing water block damage and reducing gas production. Reservoirs with high $R_{c/r}$ have larger displacement pressure, which can effectively overcome the resistance of liquid migration in pores, thereby reducing the damage of the water block. Therefore, different reservoir conditions need to match the appropriate fracturing scale. This study can provide guidance for the optimal design of hydraulic fracturing parameters for CBM wells.

Keywords: coalbed methane; hydraulic fracturing; fracturing fluid volume; analysis of variance; water block

1. Introduction

Hydraulic fracturing is one of the important reservoir stimulations of coalbed methane (CBM) wells, and its main purpose is to form efficient conductivity fractures and improve coal seam permeability [1,2]. However, while hydraulic fracturing improves the permeability of coal seams, tons of fracturing fluid are injected into the coal seam [3,4], and a large amount of fracturing fluid is leaked into the coal reservoir, which also brings many adverse effects to the development of coalbed methane, such as water block damage [5–8], clay swelling [9,10], and consequently increased difficulty in methane desorption, etc. [11,12].

In view of the adverse effects of fracturing fluid leakage, a large number of scientific research and experiments have been carried out on the influencing factors of the leakage. Yuan et al. [4] and Chang et al. [10] studied the self-absorption process and microscopic migration mechanism of coal reservoirs after hydraulic fracturing and evaluated the effect of this process on permeability. Wang et al. [13] expounded the influence of the wettability of coal on the irreducible water content from a microscopic point of view, and discussed the effective flowback of fracturing fluid after hydraulic fracturing. Meng et al. [14] studied the effects of effective stress, porosity, permeability, fracturing fluid viscosity, and reservoir pressure on the fracturing fluid leakage factor, and proposed a model that takes into account stress-sensitivity effects to estimate the overall filter leakage factor. Through numerical simulation, Guo et al. [15] showed that fracture geometry, reservoir characteristics, pressure conditions, and temperature have significant effects on fracturing fluid leakage. Guo et al. [16] observed that natural fractures play a dominant role in the leak of fracturing fluid, and the wider the opening of natural fractures, the greater the leak of fracturing fluid. From the perspective of fracturing engineering, Wu et al. [17] showed through experimental results that increasing the injection pressure will increase the fracturing fluid loss; true triaxial hydraulic fracturing experiments showed that high injection rates can cause a large amount of fracturing fluid to leak along the bedding direction [18]. However, the unreasonable fracturing scale design may also be the root cause for fracturing fluid leakage, which has rarely attracted the attention of scholars.

The design of FFV needs to consider different reservoir conditions. The appropriate fracturing fluid scale can not only achieve the maximum fracture-forming effect, and improve the permeability, but also reduce the reservoir damage caused by the leakage. At present, the optimal design of hydraulic fracturing parameters mainly uses numerical simulation methods to adjust the number, length, spacing, and conductivity of fractures, so as to determine the amount of fracturing fluid [18–21]. Usually, the numerical simulation method does not fully consider the flowback of fracturing fluid after hydraulic fracturing. When the flowback effect is not good, the fluid leakage will cause damage to the reservoir, especially related to the impact of fluid migration in the reservoir. Specially, in the process of CBM development in the southern Qinshui Basin by PetroChina Huabei Oilfield Company, the scale of hydraulic fracturing experienced a change from small to large, and the amount of fracturing fluid increased from 300 m^3 to 1000 m^3, as shown in Table 1. However, it was found that the fracture length and stimulation effect did not increase with the increase of fracturing scale, but large-scale fracturing resulted in more fracturing fluid leaking. During the development process of the Zhengzhuang (ZZ) block, the differences in geological parameters within the block were not considered, and all vertical wells have adopted the same hydraulic fracturing scale, resulting in large differences in gas production of single wells in the block [22].

Table 1. Fracturing fluid volume development stage.

Stage	Time (Year)	Fracturing Fluid Volume (m^3)
I	2006–2007	370–420
II	2008–2009	550–600
III	2010–2011	650–700
IV	2013–2014	800–1000

In this paper, we aim to better clarify whether different FFVs have an impact on gas production, and design the amount of fracturing fluid that matches the geological conditions. By analyzing the production data of 1238 wells in the Fanzhuang (FZ) block and ZZ block, the influence of different FFV on the gas production was studied, the FFV for optimal gas production under the different ratios of critical desorption pressure to the reservoir pressure ($R_{c/r}$) was discussed, and the effect of water block damage caused by fluid leakage on gas production was clarified. The research results can provide some insights into the optimal design of hydraulic fracturing parameters for CBM reservoirs.

2. Geological Setting

The Qinshui Basin, located in southeastern Shanxi Province (Figure 1a), is a typical example of the successful development of high-rank coal in China [23]. The FZ block and ZZ block are located in the southern Qinshui Basin (Figure 1b). The study area consists of the Pennsylvanian Benxi (C$_2$b) and Taiyuan (C$_2$t) Formations, the Permian Shanxi

(P$_1$s), Xiashihezi (P$_1$x), Shangshihezi (P$_2$s) and Shiqianfeng (P$_2$sh) Formations, and Triassic deposits. The main coal-bearing strata are the Shanxi and Taiyuan Formations, and the No.3 coal seam of the Shanxi Formation is stably distributed in the whole area and is the main layer for CBM development in the study area [24–26].

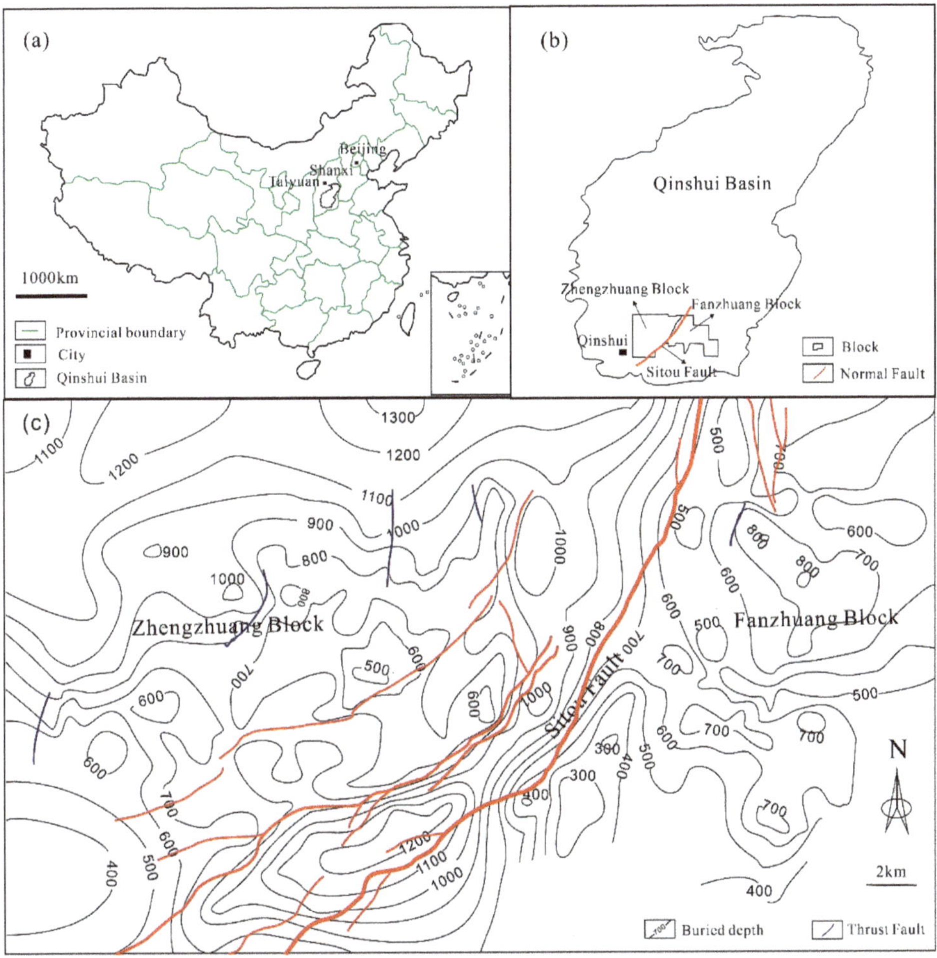

Figure 1. (**a**) Location of the Qinshui Basin in China; (**b**) location of development block in the southern Qinshui Basin; (**c**) the buried depth of the No.3 coal seam.

The FZ block and ZZ block are bounded by the Sitou Fault. The ZZ block is situated west of the fault, and the FZ block is located east of the fault [27–29]. The stratigraphic structure of the study area is relatively complex. Local folds and faults are relatively developed, and the regional structural form is mainly distributed in the north-northeast, and the stratigraphic dip is 3° to 8°. The thickness of the No. 3 coal seam ranges from 5 to 7 m, and its burial depth varies between 300 and 1200 m (Figure 1c). The vitrinite reflectance ($R_{o,max}$, %) varies between 3.1% and 3.9%, and the gas content is between 14 and 30 m^3/t. The reservoir permeability is generally lower than 1 mD, with an average of 0.27 mD [30–32].

3. Methods

In order to analyze the influence of FFV on gas production in the southern Qinshui Basin, the production data of 1238 CBM wells in FZ block and ZZ block since 2006 were collected and analyzed. These wells all use conventional hydraulic fracturing fluid. The fracturing fluid is potassium chloride solution with a concentration of 1%, and the FFV is between 200–1000 m^3. After the fracturing operation, the well is shut in, and the fracturing fluid is almost completely leaked into the reservoir. The basic data collected include FFV, critical desorption pressure, reservoir pressure, $R_{c/r}$, and peak gas production (PGP). Among them, the critical desorption pressure is the bottom hole flow pressure at the initial gas desorption during the CBM drainage process; the $R_{c/r}$ is the ratio of the critical desorption pressure to the reservoir pressure; the PGP is the maximum daily production after the first hydraulic fracturing stimulation.

In this work, one-way analysis of variance (ANOVA) was used to analyze the influence of FFV and $R_{c/r}$ on gas production. First, as many as possible FFV and $R_{c/r}$ are grouped, and then the least-significant difference (LSD) method is used to test the significant differences between the different groups, and the adjacent and insignificant groups are merged [33]. The grouping level of FFV and $R_{c/r}$ is actually obtained, which is convenient for multiway ANOVA. Then, the multiway ANOVA on the significant effect of gas production is carried out using the $R_{c/r}$ and FFV grouping level determined after one-way ANOVA, and then the optimal combination that beneficial to gas production is found [34].

In this work, SPSS software is used for ANOVA, the system default significance level α is set at 0.05 and compared with *p*-value of the test statistic. If the *p*-value < 0.05, it is considered that the different grouping levels of the independent variable have a significant impact on the dependent variable. F ratio is the test statistic, the F ratio is the between-group variance divided by the within-group variance in a data set. If F > 1, there are statistically significant differences between groups, a high F ratio indicates the greater likelihood of statistically significant differences between groups [34].

4. Results

4.1. One-Way ANOVA

Table 2 shows the results of one-way ANOVA on the PGP when the FFV is divided into 8 groups (200–300 m^3, 300–400 m^3, 400–500 m^3, 500–600 m^3, 600–700 m^3, 700–800 m^3, 800–900 m^3, and 900–1000 m^3). The results show that the amount of fracturing fluid has a significant effect on PGP (F = 20.35, *p* = 0.000). Analysis of the pairwise comparison between different groups by the LSD method shows that the difference between the 300–400 m^3 and 400–500 m^3 groups is not significant, and they are combined into a group of 300–500 m^3. Similarly, the 700–800 m^3, 800–900 m^3, and 900–1000 m^3 were combined into a group of 700–1000 m^3. The differences between other adjacent groups are significant, and the results show that the effects of different FFV groups on PGP are significantly different. The average PGP when the FFV is divided into 8 groups is shown in Figure 2a.

Table 2. One-way ANOVA results of 8 groups of FFV.

Source	Sum of Squares	df	Mean Square	F	*p*-Value
Between-group variance	6.98×10^8	7	9.97×10^7	20.35	0.000
Within-group variance	6.03×10^9	1230	4.9×10^6		
Total	6.72×10^9	1237			

df: The degree of freedom (df) of the statistic.

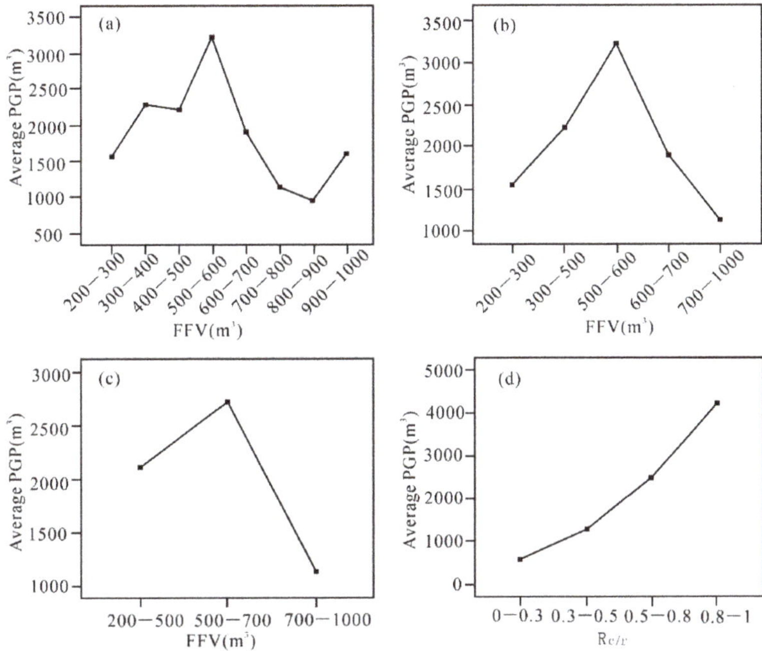

Figure 2. (**a**) Average PGP when FFV is divided into 8 groups; (**b**) average PGP when FFV is divided into 5 groups; (**c**) average PGP when FFV is divided into 3 groups; (**d**) average PGP when $R_{c/r}$ is divided into 4 groups.

Table 3 shows the results of one-way ANOVA on the PGP when the FFV is divided into 5 groups (200–300 m^3, 300–500 m^3, 500–600 m^3, 600–700 m^3, and 700–1000 m^3). The results show that the amount of fracturing fluid has a significant effect on the PGP (F = 35.1, p = 0.000). Figure 2b shows that with the increase of FFV, the average PGP first increases and then decreases. When the FFV is 500–600 m^3, the average PGP is the largest, and when the FFV exceeds 500–600 m^3, the average PGP shows a rapid downward trend.

Table 3. One-way ANOVA results of 5 groups of FFV.

Source	Sum of Squares	df	Mean Square	F	*p*-Value
Between-group variance	6.86×10^8	4	1.72×10^8	35.1	0.000
Within-group variance	6.04×10^9	1233	4.9×10^6		
Total	6.72×10^9	1237			

df: The degree of freedom (df) of the statistic.

Table 4 shows the results of one-way ANOVA on the PGP when the FFV is divided into 3 groups (200–500 m^3, 500–700 m^3, and 700–1000 m^3). The results show that the amount of fracturing fluid has a significant effect on PGP (F = 54.6, p = 0.000). The LSD method analyzes the pairwise comparison between the three different grouping levels, and the results show that the differences between the groups are significant. It can be known from Figure 2c that when the FFV is 200–500 m^3, the average PGP is 2136 m^3; when the FFV is 500–700 m^3, the average PGP is 2795 m^3; and when the FFV is 700–1000 m^3, the average PGP dropped rapidly to only 1121 m^3.

Table 4. One-way ANOVA results of 3 groups of FFV.

Source	Sum of Squares	df	Mean Square	F	*p*-Value
Between-group variance	5.46×10^8	2	2.73×10^8	54.6	0.000
Within-group variance	6.18×10^9	1235	5×10^6		
Total	6.72×10^9	1237			

df: The degree of freedom (df) of the statistic.

The one-way ANOVA process of $R_{c/r}$ on gas production is shown in Table 5, which shows that $R_{c/r}$ has a significant impact on gas production when $R_{c/r}$ is divided into 10, 7, and 6 groups. Table 6 shows the results of one-way ANOVA on PGP when the $R_{c/r}$ is divided into 4 groups (0–0.3, 0.3–0.5, 0.5–0.8 and 0.8–1). The results showed that the $R_{c/r}$ had a significant impact on the PGP (F = 111.53, p = 0.000). The results of pairwise comparison between different groups by the LSD method showed that the differences in PGP between the four different grouping levels were significant. Figure 2d shows that the average PGP is positively correlated with the $R_{c/r}$.

Table 5. F and *p*-value of one-way ANOVA for different groups of $R_{c/r}$.

Variable	Number of Grouping Levels	Grouping Level	F	*p*-Value
$R_{c/r}$	10	0–0.1, 0.1–0.2, 0.2–0.3, 0.3–0.4, 0.4–0.5, 0.5–0.6, 0.6–0.7, 0.7–0.8, 0.8–0.9, 0.9–1	42.2	0.000
	7	0–0.3, 0.3–0.4, 0.4–0.5, 0.5–0.6, 0.6–0.8, 0.8–0.9, 0.9–1	64.9	0.000
	6	0–0.3, 0.3–0.4, 0.4–0.5, 0.5–0.8, 0.8–0.9, 0.9–1	77.1	0.000

Table 6. One-way ANOVA results of 4 groups of $R_{c/r}$.

Source	Sum of Squares	df	Mean Square	F	*p*-Value
Between-group variance	1.48×10^9	3	4.74×10^8	111.53	0.000
Within-group variance	5.24×10^9	1234	4.25×10^6		
Total	6.72×10^9	1237			

df: The degree of freedom (df) of the statistic.

One-way ANOVA shows that both the FFV and $R_{c/r}$ have a significant impact on the gas production. Combined with the actual situation, the FFV is divided into 3 groups and the $R_{c/r}$ is divided into 4 groups for multiway ANOVA, as shown in Table 7.

Table 7. Multiway ANOVA variable grouping level table.

Variable	Number of Grouping Levels	Grouping Level	Number of Wells	Total Number of Wells
$R_{c/r}$	4	0–0.3	231	1238
		0.3–0.5	484	
		0.5–0.8	381	
		0.8–1	142	
FFV	3	200–500 m^3	459	1238
		500–700 m^3	259	
		700–1000 m^3	520	

df: The degree of freedom (df) of the statistic.

4.2. Multiway ANOVA

The Levene's test results [F (11, 1226) = 23.76, p = 0.000] of the multiway ANOVA of the $R_{c/r}$ and FFV can show that the overall variance of the samples in each group is

homogeneous, which meets the precondition of the variance test. Table 8 shows that the model used for the multiway ANOVA was statistically significant (F = 40.2, p = 0.000). The interaction between $R_{c/r}$ and FFV had a very significant impact on PGP (F = 7.42, p = 0.000).

Table 8. Multiway ANOVA results of Rc/r and FFV.

Source	Sum of Squares	df	Mean Square	F	p-Value
Correction model	1.78×10^9	11	1.62×10^8	40.2	0.000
$R_{c/r}$	1.05×10^9	3	3.49×10^8	86.6	0.000
FFV	5.68×10^7	2	2.84×10^7	7.05	0.001
$R_{c/r}$:FFV	1.79×10^8	6	2.99×10^7	7.42	0.000
Error	4.94×10^9	1226	4.03×10^6		
Total	6.72×10^9	1237			

df: The degree of freedom (df) of the statistic.

The multiway ANOVA shows that when the FFV is constant, the larger the $R_{c/r}$, the better the gas production, and the $R_{c/r}$ have a significant contribution to the gas production. When the $R_{c/r}$ is constant, the amount of fracturing fluid is different, and the gas production is also different. The specific performance is as follows: (a) when $R_{c/r} < 0.5$, the gas production is negatively correlated with the amount of fracturing fluid. when $R_{c/r}$ is 0–0.3, the FFV is 700–1000 m^3 and the gas production decreases rapidly; when $R_{c/r}$ is 0.3–0.5, the FFV is 500–700 m^3 and the gas production decreases rapidly. (b) when $R_{c/r} > 0.5$, the gas production increases first and then decreases with the increase of the FFV, but there are differences. When $R_{c/r}$ is 0.5–0.8, the gas production when the FFV is 700–1000 m^3 is less than that when the FFV is 200–500 m^3. When $R_{c/r}$ is 0.8–1, the gas production when the FFV is 700–1000 m^3 is greater than that when the FFV is 200–500 m^3. In short, the gas production decreases when the FFV is 700–1000 m^3, indicating that excessive FFV is not conducive to the increase of gas production (Figure 3).

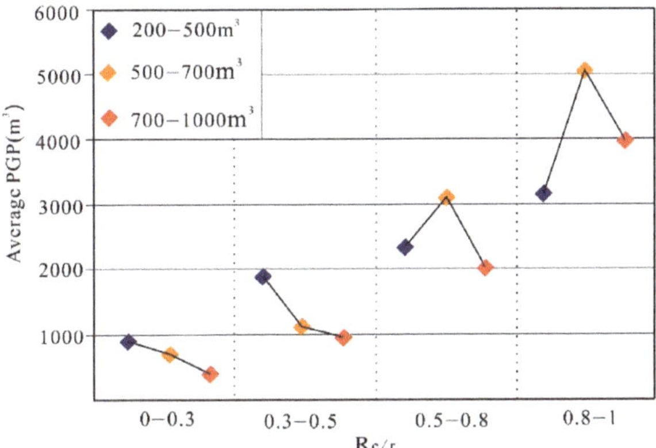

Figure 3. The average peak gas production when $R_{c/r}$ interacts with FFV.

5. Discussion

5.1. Intrusion and Retention of Fracturing Fluid

The process of hydraulic fracturing of CBM wells is the process of intrusion of fracturing fluid into pores and fractures of coal seams. The schematic diagram of fluid distribution in pores and fractures before hydraulic fracturing is shown in Figure 4a. The water intrusion process is mainly affected by injection pressure (P_d), imbibition capillary force (P_c), viscous resistance (P_n), fluid resistance (P_f), and gas pressure (P_g) in the coal seam. It is generally considered that P_d and P_c are the main driving forces [35,36]. When the coal reservoir is saturated with gas and contains more free methane gas, P_g is also a non-negligible resistance to prevent water migration in pores [37,38]. Therefore, for the fluid migrating in the pores and fractures of coal seams during the water invasion process, the pressure difference (ΔP_i) across the pores is:

$$\Delta P_i = P_d + P_c - P_n - P_f - P_g \tag{1}$$

$$P_c = \frac{2\delta \cos \theta}{r} \tag{2}$$

where σ is the interfacial tension between the solution and the air, N/m; θ is the contact angle between the solution and coal, (°); r is the radius of the pore, m.

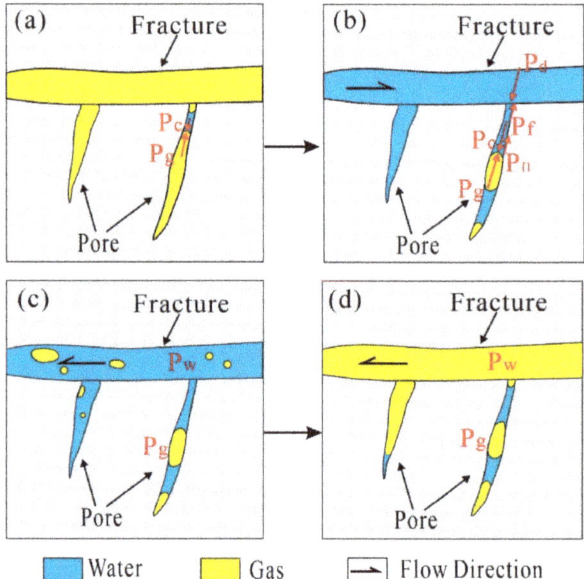

Figure 4. (**a**) Fluid distribution in pores and fractures before hydraulic fracturing; (**b**) fluid distribution in pores and fractures after hydraulic fracturing; (**c**) fluid distribution in pores and fractures at the early stage of drainage; (**d**) fluid distribution in pores and fractures at the later stage of drainage.

When ΔP_i is >0, water intrusion occurs. The water retained in the pores may fill the pores, or may form multi-level and intermittent water columns (Figure 4b).

The water intrusion experiments in coal pillars show that water intrusion occurs simultaneously in micropore-transition pores, mesopores, macropores and fractures, and the water intrusion rate decreases sequentially. The speed of water intrusion in the micropore-transition pore is mainly determined by the capillary force of imbibition; the more complex the pore structure, the smaller the degree of water intrusion, and the more difficult it is to flow back after water intrusion. The water saturation of pores and fractures increased with the increase of injection time and inlet pressure during the water invasion process [39].

In the practice of hydraulic fracturing, with the increase of the fracturing scale, the injection rate and pressure need to be increased accordingly. In this way, the amount of fracturing fluid invading into the pores also increases, and the radius of the pores that can be invaded is smaller, and more fracturing fluid enters the complex pores and micropores. Therefore, if the scale of hydraulic fracturing is too large, the more fracturing fluid that is leaked and retained in the pores and fractures, and the residual fracturing fluid will affect the gas production [40]. As shown in Figure 3, when the amount of fracturing fluid is 700–1000 m^3, the average PGP decreases compared with that when the amount of fracturing fluid is 500–700 m^3, and more external liquid stays in the pores and fractures of the coal.

5.2. Fluid Migration and Water Block Damage during Drainage

At the early stage of CBM well drainage, there is saturated water in the pores and fissures. As the water in the fracture is drained first, the fluid pressure in the fracture will decrease and the gas begins to desorb (Figure 4c). After the water in the fracture is drained out, some of the water in the pore does not migrate with the water in the fracture, but stays in the pore (Figure 4d). According to the principle of gas-liquid two-phase fluid flow, it can be known that the fluid in the pore mainly considers the two-phase flow driven by the pressure difference [41]. The pressure difference for liquid column migration in the pore is [13,42]:

$$\Delta P_o = P_g - P_c - P_w - P_f - G \tag{3}$$

where P_g is the gas pressure in the pore; P_w is the fluid pressure in the fracture; and G is the gravity of the liquid column in the pore.

P_g is the main driving force for the liquid column migration in the pore, and P_c and P_w are the main resistances. When the pore radius is small enough and the liquid column is short enough, P_f and G can be ignored. When $\Delta P_o = 0$, the fluid in the pore does not migrate (Figure 5a). As the fluid in the fracture migrates out, P_w will decrease, the pressure drop will be transferred to the pore, and part of the adsorbed gas will be desorbed from the pore, and P_g will increase. When $\Delta P_o > 0$, the liquid column migrates to the fracture. At this time, the fluid pressure balance in the pore is destroyed, and the gas at the bottom of the pore will push the liquid column at the bottom upward until a new balance is reached (Figure 5b). With the progress of drainage, the pressure drop is effectively transferred to the internal pores, and more gas is desorbed. When the flow resistance of the liquid column can be overcome, the gas will break through the constraints of the liquid and migrate out. At this time, the pores and fractures are fully connected (Figure 5c). In this process, the part of the liquid column that cannot overcome its flow resistance is bound in the pores, blocking the pores, affecting the migration of gas, and forming water block damage.

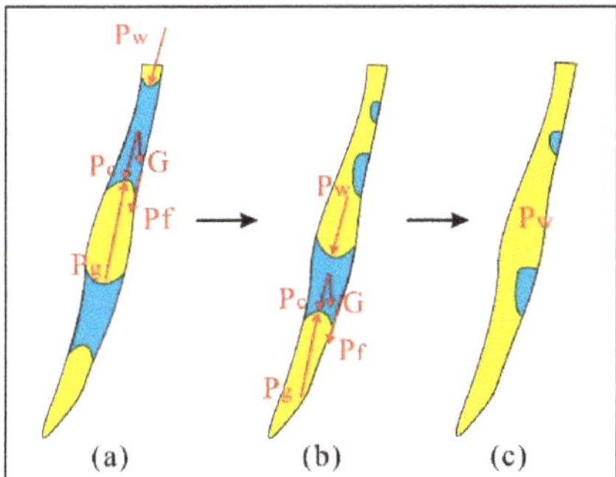

Figure 5. (a) Schematic diagram of the force when the fluid in the pores does not migrate; (b) schematic diagram of the fluid in the pores in a new balance state; (c) schematic diagram when pores and fractures are fully connected.

The gas flooding experiment also showed that the water block damage of macropores and fractures can be relieved, and the water block of mesopores can be partially relieved, while the water in micropores and transition pores was difficult to displace [39,43,44]. In actual production, a large amount of fracturing fluid invades into the pores, and there is no additional driving force during the drainage process, and it will be more difficult for the fracturing fluid to be completely discharged from the micropores and transition pores. Therefore, this part of water must rely on the driving force of the gas in the pores to be discharged.

5.3. Mechanism of FFV Affecting Gas Production

It can be seen from the above analysis that water block mainly comes in two ways: (1) during the fracturing process, a large amount of fracturing fluid intrudes into the pores and fractures, and is trapped by capillary force; (2) in the process of drainage, the driving force of the gas in the pore is not enough to overcome the resistance of its migration, and the fracturing fluid retained in the pore cannot be discharged back. According to the Hagen–Poiseuille law, the volume of the fracturing fluid discharged from the pores against the capillary resistance is [45]:

$$Q = \frac{\pi r^4 \Delta P_o}{8 \mu L} \tag{4}$$

where Q is the volume of fracturing fluid discharged, m^3; r is the radius of the pore, m; μ is the dynamic viscosity, Pa s; L is the length of the liquid column, m.

Take the derivative of Equation (4) as

$$\frac{dL}{dt} = \frac{r^2 \Delta P_o}{8 \mu L} \tag{5}$$

By the integral of Equation (5), it can be obtained that the time (t) for the liquid column of length (L) to flow back from the pores is

$$t = \frac{4 \mu L^2}{\Delta P_o r^2} \tag{6}$$

Substitute Equation (3) into Equation (6):

$$t = \frac{4 \mu L^2}{(P_g - P_c - P_w - P_f - G) r^2} \tag{7}$$

When the scale of hydraulic fracturing fluid increases, the amount of invading fluid in pores increases, the length of the liquid column (L) increases, and the radius (r) of the pores that can be invaded becomes smaller; at the same time, the P and G of the liquid column will have to be considered, increasing the resistance and time for the fluid to move out, making it easier to cause water block [37].

According to the analysis of the microseismic fracture monitoring report of CBM wells in the study area, the FFV has no significant effect on the length of the fracture, and the length of the fracture does not increase with the increase of the FFV (Figure 6). The correlation between gas production and fracture length is also not obvious (Figure 7). It can be seen that increasing the scale of fracturing has not always brought positive effects on gas production [22]. From the one-way ANOVA of FFV and PGP (Figure 2), it can be seen that with the increase of FFV, the PGP increases first and then decreases. Within a certain scale, increasing the amount of fracturing fluid has a positive effect on gas production. When it exceeds a certain scale, it will have a negative effect on gas production. This is because excess fracturing fluid does not play a role in creating fractures and increasing reservoir connectivity. The excess fracturing fluid is leaked into the coal reservoir or surrounding rock along the pore and fracture channels, and the positive effect is not as great as the negative effect of water block caused by excessive fracturing fluid staying in the reservoir [46]. Therefore, the negative impact of FFV on gas production is mainly reflected in the water block damage caused to the reservoir.

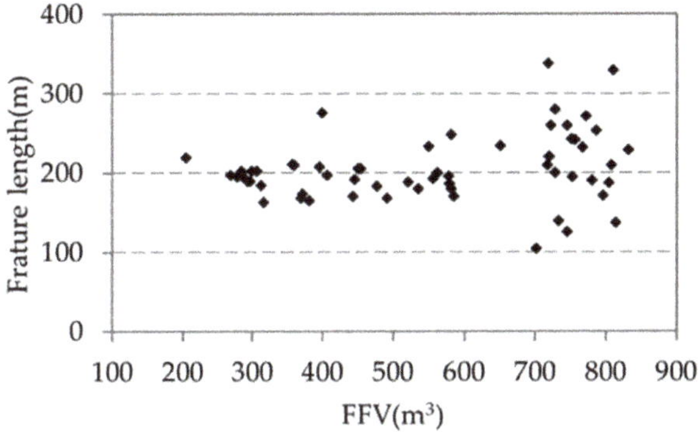

Figure 6. Diagram of FFV and hydraulic fracturing fracture length.

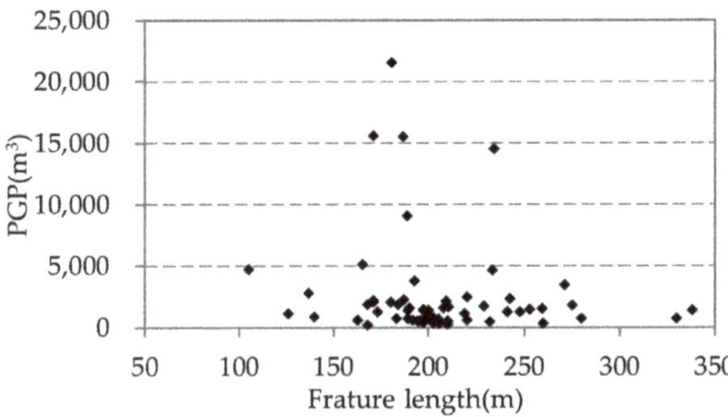

Figure 7. Diagram of hydraulic fracturing fracture length and PGP.

It can be seen from Equation (7) that the greater the ΔP_o, the greater the fluid resistance that can be overcome. Therefore, under different reservoir conditions, the degree of difficulty in releasing the water block is also different. Lyu et al. [38] and Lu et al. [42] showed that water block did not occur when the differential pressure driving force was greater than the resistance. The larger the $R_{c/r}$ value, the better the fluid drainage index of the reservoir, which was more conducive to the desorption of gas, resulting in high production [47]. The $R_{c/r}$ value can represent the gas saturation to a certain extent. Higher gas saturation leads to higher desorption pressure and is beneficial for early gas production, leading to greater total gas production [48,49]. It can be seen that the larger the $R_{c/r}$ value, the easier the gas desorption, the greater the pressure of gas in the pore, and the greater the driving force of liquid migration in the pore. Therefore, when the $R_{c/r}$ increases, it becomes easier for the water in the pores to be displaced out.

As shown in Figure 3, when $R_{c/r} > 0.5$, the gas production with the FFV of 500–700 m^3 is better than that with the FFV of 200–500 m^3, indicating that the positive effect of increasing the FFV on the reservoir is greater than the water block damage to the reservoir. When the amount of fracturing fluid increases to 700–1000 m^3, the gas production decreases, indicating that the damage of the water block caused by increasing the volume of fracturing fluid is greater than the positive effect caused by increasing the volume of fracturing fluid. In particular, when $R_{c/r}$ is 0.5–0.8, the gas production with the FFV of 700–1000 m^3 is smaller than that with the FFV of 200–500 m^3, implying that the negative effect of excessively increasing the amount of fracturing fluid is greater, and more water blocks are not released. When $R_{c/r}$ is 0.8–1, the gas production with the FFV of 700–1000 m^3 is higher than that of the FFV of 200–500 m^3, indicating that the positive effect brought by the large FFV is greater than the negative effect. The water block is easier to be eliminated when the

$R_{c/r}$ is 0.8–1 than when the $R_{c/r}$ is 0.5–0.8. When $R_{c/r} < 0.5$, the gas production will be less when the FFV is increased, which indicates that the fracturing fluid enters the pores and is difficult to be displaced, which is more likely to cause water block damage to the reservoir. Therefore, the smaller the $R_{c/r}$ value, the larger the FFV, and the more serious the water block damage.

5.4. The Uncertainty or Limitation of This Study

There are many factors affecting the production of CBM [32,50–53]. This paper illustrates that the amount of fracturing fluid is also one of the factors affecting the production of gas through the ANOVA of production data. This paper selects the $R_{c/r}$ and FFV for multivariate ANOVA, and proposes that the design of FFV can be based on the $R_{c/r}$ in the study area. The combination of other geological factors and FFV may also affect gas production and multivariate ANOVA of other geological parameters, and FFV may be attempted later. In this paper, the coal samples in the study area are not used for the analysis of water intrusion, drainage, and water block in pores and fractures, and the experiment of Li et al. [40] is cited for illustration. The data in this paper are from the FZ block and ZZ block in Qinshui Basin, which is a high-rank coal, so the applicability of the optimal combination and the reference of FFV parameter design are only applicable to this block.

6. Conclusions

(1) One-way ANOVA shows that FFV has a significant effect on gas production. In particular, with the increase of FFV, the gas production increases first and then decreases, and the best gas production is when the FFV is 500–600 m^3 in the study area.

(2) When the $R_{c/r} < 0.5$ (0–0.3 and 0.3–0.5), the gas production is negatively correlated with the FFV. When the $R_{c/r} > 0.5$ (0.5–0.8 and 0.8–1), the gas production increases first and then decreases with the increase of FFV. The best combination for gas production is the $R_{c/r}$ of 0.8–1 and the FFV of 500–700 m^3.

(3) It is found that too much injected fracturing fluid will increase the fluid leakage and lead to the water block damage of coal reservoir, and the increase of the $R_{c/r}$ is conducive to removing the water block damage. Therefore, it is necessary to adapt the optimal amount of fracturing fluid according to the condition of the $R_{c/r}$ in the study area, so as to achieve the best fracturing effect.

Author Contributions: Methodology, writing—original draft, W.C. (Wenwen Chen); data curation, investigation, F.Q. and W.C. (Weiwei Chao); data curation, resources, M.T. and W.C. (Wei Chen); revised and edited the manuscript, X.W. and S.H. All authors have read and agreed to the published version of the manuscript.

Funding: This research was funded by National Science Foundation of China (No. 41972184, No. 42262022 and No. 41902177) and Jiangxi Provincial Natural Science Foundation (grant number 20212BAB214030).

Acknowledgments: The authors would like to give their sincere thanks to the teachers in the department for their comments on the revision of the article, and thanks to the former colleagues of PetroChina Huabei Oilfield Company for their valuable comments.

Conflicts of Interest: The authors declare no conflict of interest.

References

1. Xu, B.; Li, X.; Haghighi, M.; Ren, W.; Du, X.; Chen, D.; Zhao, Y. Optimization of hydraulically fractured well configuration in anisotropic coal-bed methane reservoirs. *Fuel* **2013**, *107*, 859–865. [CrossRef]
2. Yuan, B.; Wood, D.A.; Yu, W. Stimulation and hydraulic fracturing technology in natural gas reservoirs: Theory and case studies (2012–2015). *J. Nat. Gas Sci. Eng.* **2015**, *26*, 1414–1421. [CrossRef]
3. Yan, Q.; Lemanski, C.; Karpyn, Z.T.; Ayala, L.F. Experimental investigation of shale gas production impairment due to fracturing fluid migration during shut-in time. *J. Nat. Gas Sci. Eng.* **2015**, *24*, 99–105. [CrossRef]
4. Yuan, X.; Yao, Y.; Liu, D.; Pan, Z. Spontaneous imbibition in coal: Experimental and model analysis. *J. Nat. Gas Sci. Eng.* **2019**, *67*, 108–121. [CrossRef]

5. Naik, S.; You, Z.; Bedrikovetsky, P. Rate enhancement in unconventional gas reservoirs by wettability alteration. *J. Nat. Gas Sci. Eng.* **2015**, *26*, 1573–1584. [CrossRef]

6. Su, X.; Wang, Q.; Song, J.; Chen, P.; Yao, S.; Hong, J.; Zhou, F. Experimental study of water blocking damage on coal. *J. Pet. Sci. Eng.* **2017**, *156*, 654–661. [CrossRef]

7. Huang, Q.; Liu, S.; Cheng, W.; Wang, G. Fracture permeability damage and recovery behaviors with fracturing fluid treatment of coal: An experimental study. *Fuel* **2020**, *282*, 118809. [CrossRef]

8. Chang, Y.; Yao, Y.; Liu, D.; Liu, Y.; Cui, C.; Wu, H. Behavior and mechanism of water imbibition and its influence on gas permeability during hydro-fracturing of a coalbed methane reservoir. *J. Pet. Sci. Eng.* **2022**, *208*, 109745. [CrossRef]

9. Bahrami, H.; Rezaee, R.; Clennell, B. Water blocking damage in hydraulically fractured tight sand gas reservoirs: An example from Perth Basin, Western Australia. *J. Pet. Sci. Eng.* **2012**, *88–89*, 100–106. [CrossRef]

10. Yang, S.; Wen, G.; Yan, F.; Li, H.; Liu, Y.; Wu, W. Swelling characteristics and permeability evolution of anthracite coal containing expansive clay under water-saturated conditions. *Fuel* **2020**, *279*, 118501. [CrossRef]

11. Li, X.; Kang, Y. Effect of fracturing fluid immersion on methane adsorption/desorption of coal. *J. Nat. Gas Sci. Eng.* **2016**, *34*, 449–457. [CrossRef]

12. Lu, W.; Huang, B.; Zhao, X. A review of recent research and development of the effect of hydraulic fracturing on gas adsorption and desorption in coal seams. *Adsorpt. Sci. Technol.* **2019**, *37*, 509–529. [CrossRef]

13. Wang, Z.; Liu, S.; Qin, Y. Coal wettability in coalbed methane production: A critical review. *Fuel* **2021**, *303*, 121277. [CrossRef]

14. Meng, Y.; Li, Z.; Lai, F. Evaluating the filtration property of fracturing fluid and fracture conductivity of coalbed methane wells considering the stress-sensitivity effects. *J. Nat. Gas Sci. Eng.* **2020**, *80*, 103379. [CrossRef]

15. Guo, J.; Liu, Y. A comprehensive model for simulating fracturing fluid leakoff in natural fractures. *J. Nat. Gas Sci. Eng.* **2014**, *21*, 977–985. [CrossRef]

16. Guo, J.; Liu, Y. Opening of natural fracture and its effect on leakoff behavior in fractured gas reservoirs. *J. Nat. Gas Sci. Eng.* **2014**, *18*, 324–328. [CrossRef]

17. Wu, C.; Zhang, X.; Wang, M.; Zhou, L.; Jiang, W. Physical simulation study on the hydraulic fracture propagation of coalbed methane well. *J. Appl. Geophys.* **2018**, *150*, 244–253. [CrossRef]

18. Jiang, T.; Zhang, J.; Wu, H. Experimental and numerical study on hydraulic fracture propagation in coalbed methane reservoir. *J. Nat. Gas Sci. Eng.* **2016**, *35*, 455–467. [CrossRef]

19. Jahandideh, A.; Jafarpour, B. Optimization of hydraulic fracturing design under spatially variable shale fracability. *J. Pet. Sci. Eng.* **2016**, *138*, 174–188. [CrossRef]

20. Zhao, J.; Zhao, J.; Hu, Y.; Zhang, S.; Huang, T.; Liu, X. Numerical simulation of multistage fracturing optimization and application in coalbed methane horizontal wells. *Eng. Fract. Mech.* **2020**, *223*, 106738. [CrossRef]

21. Song, H.; Du, S.; Yang, J.; Zhao, Y.; Yu, M. Evaluation of hydraulic fracturing effect on coalbed methane reservoir based on deep learning method considering physical constraints. *J. Pet. Sci. Eng.* **2022**, *212*, 110360. [CrossRef]

22. Zhu, Q.; Zuo, Y.; Yang, Y. How to solve the technical problems in the CBM development: A case study of a CBM gas reservoir in the southern Qinshui Basin. *Nat. Gas Ind.* **2015**, *35*, 106–109. [CrossRef]

23. Chen, S.; Tang, D.; Tao, S.; Xu, H.; Li, S.; Zhao, J.; Cui, Y.; Li, Z. Characteristics of in-situ stress distribution and its significance on the coalbed methane (CBM) development in Fanzhuang-Zhengzhuang Block, Southern Qinshui Basin, China. *J. Pet. Sci. Eng.* **2018**, *161*, 108–120. [CrossRef]

24. Su, X.; Li, X.; Li, S.; Zhao, M.; Song, Y. Geology of coalbed methane reservoirs in the Southeast Qinshui Basin of China. *Int. J. Coal Geol.* **2005**, *62*, 197–210. [CrossRef]

25. Zhang, J.; Liu, D.; Cai, Y.; Pan, Z.; Yao, Y.; Wang, Y. Geological and hydrological controls on the accumulation of coalbed methane within the No. 3 coal seam of the southern Qinshui Basin. *Int. J. Coal Geol.* **2017**, *182*, 94–111. [CrossRef]

26. Han, W.; Wang, Y.; Li, Y.; Ni, X.; Wu, X.; Wu, P.; Zhao, S. Recognizing fracture distribution within the coalbed methane reservoir and its implication for hydraulic fracturing: A method combining field observation, well logging, and micro-seismic detection. *J. Nat. Gas Sci. Eng.* **2021**, *92*, 103986. [CrossRef]

27. Chen, Y.; Liu, D.; Yao, Y.; Cai, Y.; Chen, L. Dynamic permeability change during coalbed methane production and its controlling factors. *J. Nat. Gas Sci. Eng.* **2015**, *25*, 335–346. [CrossRef]

28. Zhao, X.; Yang, Y.; Sun, F.; Wang, B.; Zuo, Y.; Li, M.; Shen, J.; Mu, F. Enrichment mechanism and exploration and development technologies of high coal rank coalbed methane in south Qinshui Basin, Shanxi Province. *Pet. Explor. Dev.* **2016**, *43*, 332–339. [CrossRef]

29. Shi, J.; Zeng, L.; Zhao, X.; Zhao, Y.; Wang, J. Characteristics of natural fractures in the upper Paleozoic coal bearing strata in the southern Qinshui Basin, China: Implications for coalbed methane (CBM) development. *Mar. Pet. Geol.* **2020**, *113*, 104152. [CrossRef]

30. Cai, Y.; Liu, D.; Yao, Y.; Li, J.; Qin, Y. Geological controls on prediction of coalbed methane of No. 3 coal seam in Southern Qinshui Basin, North China. *Int. J. Coal Geol.* **2011**, *88*, 101–112. [CrossRef]

31. Zhang, J.; Liu, D.; Cai, Y.; Yao, Y.; Ge, X. Carbon isotopic characteristics of CH4 and its significance to the gas performance of coal reservoirs in the Zhengzhuang area, Southern Qinshui Basin, North China. *J. Nat. Gas Sci. Eng.* **2018**, *58*, 135–151. [CrossRef]

32. Shen, J.; Qin, Y.; Li, Y.; Yang, Y.; Ju, W.; Yang, C.; Wang, G. In situ stress field in the FZ Block of Qinshui Basin, China: Implications for the permeability and coalbed methane production. *J. Pet. Sci. Eng.* **2018**, *170*, 744–754. [CrossRef]

33. Lizasoain, L.; Joaristi, L. *Management and Data Analysis with SPSS*; Paraninfo S.A.: Madrid, Spain, 2003; p. 480.
34. Harris, J.; Sheean, P.; Gleason, P.; Bruemmer, B.; Boushey, C. Publishing Nutrition Research: A Review of Multivariate Techniques—Part 2: Analysis of Variance. *J. Acad. Nutr. Diet.* **2012**, *112*, 90–98. [CrossRef]
35. Yao, Y.; Liu, D.; Liu, J.; Xie, S. Assessing the water migration and permeability of large intact bituminous and anthracite coals using NMR relaxation spectrometry. *Transp. Porous Media* **2015**, *107*, 527–542. [CrossRef]
36. Sun, X.; Yao, Y.; Liu, D.; Zhou, Y. Investigations of CO2-water wettability of coal: NMR relaxation method. *Int. J. Coal Geol.* **2018**, *188*, 38–50. [CrossRef]
37. Lyu, S.; Chen, X.; Shah, S.M.; Wu, X. Experimental study of influence of natural surfactant soybean phospholipid on wettability of high-rank coal. *Fuel* **2019**, *239*, 1–12. [CrossRef]
38. Lyu, S.; Wang, S.; Li, J.; Chen, X.; Chen, L.; Dong, Q.; Zhang, X.; Huang, P. Massive Hydraulic Fracturing to Control Gas Outbursts in Soft Coal Seams. *Rock Mech. Rock Eng.* **2022**, *55*, 1759–1776. [CrossRef]
39. Li, X.; Fu, X.; Ranjith, P.G.; Fang, Y. Retained water content after nitrogen driving water on flooding saturated high volatile bituminous coal using low-field nuclear magnetic resonance. *J. Nat. Gas Sci. Eng.* **2018**, *57*, 189–202. [CrossRef]
40. Sharma, M.; Agarwal, S. Impact of liquid loading in hydraulic fractures on well productivity. *J. Pet. Technol.* **2013**, *65*, 162–165. [CrossRef]
41. Li, X.; Shi, J.; Du, X.; Hu, A.; Chen, D.; Zhang, D. Transport mechanism of desorbed gas in coalbed methane reservoirs. *Pet. Explor. Dev.* **2012**, *39*, 203–213. [CrossRef]
42. Lu, Y.; Li, H.; Lu, J.; Shi, S.; Wang, G.; Ye, Q.; Li, R.; Zhu, X. Clean up water blocking damage in coalbed methane reservoirs by microwave heating: Laboratory studies. *Process Saf. Environ. Prot.* **2020**, *138*, 292–299. [CrossRef]
43. Connolly, P.R.J.; Vogt, S.J.; Iglauer, S.; May, E.F.; Johns, M.L. Capillary trapping quantification in sandstones using NMR relaxometry. *Water Resour. Res.* **2017**, *53*, 7917–7932. [CrossRef]
44. He, F.; Wang, J. Study on the Causes of Water Blocking Damage and Its Solutions in Gas Reservoirs with Microfluidic Technology. *Energies* **2022**, *15*, 2684. [CrossRef]
45. Liu, D.; Yao, Y.; Yuan, X.; Yang, Y. Experimental evaluation of the dynamic water-blocking effect in coalbed methane reservoir. *J. Petrol. Sci. Eng.* **2022**, *217*, 110887. [CrossRef]
46. Chen, M.; Bai, J.; Kang, Y.; Chen, Z.; You, L.; Li, X.; Liu, J.; Zhang, Y. Redistribution of fracturing fluid in shales and its impact on gas transport capacity. *J. Nat. Gas Sci. Eng.* **2020**, *86*, 103747. [CrossRef]
47. Zhao, X.; Yang, Y.; Chen, L.; Yang, Y.; Shen, J.; Chao, W.; Shao, G. Production controlling mechanism and mode of solid-fluid coupling of high rank coal reservoirs. *Acta Pet. Sinic.* **2015**, *36*, 1024–1034. [CrossRef]
48. Tao, S.; Tang, D.Z.; Xu, H.; Gao, L.J.; Fang, Y. Factors controlling high-yield coalbed methane vertical wells in the Fanzhuang Block, Southern Qinshui Basin. *Int. J. Coal Geol.* **2014**, *134*, 38–45. [CrossRef]
49. Tao, S.; Pan, Z.; Tang, S.; Chen, S. Current status and geological conditions for the applicability of CBM drilling technologies in China: A review. *Int. J. Coal Geol.* **2019**, *202*, 59–108. [CrossRef]
50. Moore, T. Coalbed methane: A review. *Int. J. Coal Geol.* **2012**, *101*, 36–81. [CrossRef]
51. Guo, Q.; Fink, R.; Littke, R.; Zieger, L. Methane sorption behaviour of coals altered by igneous intrusion, South Sumatra Basin. *Int. J. Coal Geol.* **2019**, *214*, 103250. [CrossRef]
52. Fu, H.; Yan, D.; Su, X.; Wang, J.; Li, Q.; Li, X.; Zhao, W.; Zhang, L.; Wang, X.; Li, Y. Biodegradation of early thermogenic gas and generation of secondary microbial gas in the Tieliekedong region of the northern Tarim Basin, NW China. *Int. J. Coal Geol.* **2022**, *261*, 104075. [CrossRef]
53. Qin, Y.; Moore, T.; Shen, J.; Yang, Z.; Shen, Y.; Wang, G. Resources and geology of coalbed methane in China: A review. *Int. Geol. Rev.* **2018**, *60*, 777–817. [CrossRef]

Article

Depositional Environment and Organic Matter Enrichment of Early Cambrian Niutitang Black Shales in the Upper Yangtze Region, China

Peng Xia [1,2,3,*], Fang Hao [3], Jinqiang Tian [3], Wenxi Zhou [1], Yong Fu [1,2], Chuan Guo [1,2], Zhen Yang [1], Kunjie Li [4] and Ke Wang [1,2]

[1]   Resource and Environmental Engineering College, Guizhou University, Guiyang 550025, China; gzzhouwenxi@126.com (W.Z.); byez1225@126.com (Y.F.); cguo@gzu.edu.cn (C.G.); cugbyzh@163.com (Z.Y.); kwang5@gzu.edu.cn (K.W.)

[2]   Key Laboratory of Ministry of Education for Geological Resources and Environment, Guiyang 550025, China

[3]   School of Geosciences, China University of Petroleum, Qingdao 266580, China; haofang@cup.edu.cn (F.H.); tianjingqiang@cup.edu.cn (J.T.)

[4]   Huaxin Gas Group Company Limited, Taiyuan 030026, China; marden2008@163.com

*   Correspondence: pxia@gzu.edu.cn

Citation: Xia, P.; Hao, F.; Tian, J.; Zhou, W.; Fu, Y.; Guo, C.; Yang, Z.; Li, K.; Wang, K. Depositional Environment and Organic Matter Enrichment of Early Cambrian Niutitang Black Shales in the Upper Yangtze Region, China. *Energies* **2022**, *15*, 4551. https://doi.org/10.3390/en15134551

Academic Editors: Jing Li, Yidong Cai and Lei Zhao

Received: 12 May 2022
Accepted: 15 June 2022
Published: 22 June 2022

Publisher's Note: MDPI stays neutral with regard to jurisdictional claims in published maps and institutional affiliations.

Abstract: Natural gas generation is the result of organic matter degradation under the effects of biodegradation and thermal degradation. Early Cambrian black shales in the Upper Yangtze Region are rich in organic matter and have shown great shale gas potentiality in recent years. Nevertheless, the enrichment mechanism and distribution of organic matter in these black shales between different sedimentary settings, such as intra-platform basin, slope, and deep basin, are still poorly understood. In this paper, based mainly on elemental geochemistry, a comprehensive study of the marine redox conditions, primary productivity, sedimentation rate, terrigenous input, hydrothermal activity, and water mass restrictions was conducted on the Early Cambrian Niutitang black shale in the Upper Yangtze Region. Our data showed that an intra-platform basin received a higher terrigenous input and that it deposited under more restricted conditions than the slope and deep basin settings. The primary productivity in the slope and deep basin settings was higher than that in the intra-platform basin setting. In the intra-platform basin, the productivity increased from its inner part to its margin. For the slope and deep basin settings, the high paleoproductivity generated large amounts of organic matter and its preservation was synergistically affected by the redox conditions. In contrast to the slope and deep basin, the preservation of organic matter in the inner part of the intra-platform basin was mainly controlled by redox conditions because the paleoproductivity in it was much lower than in the slope and deep basin settings. The intra-platform basin margin was the most favorable area for accumulating organic matter.

Keywords: Early Cambrian; paleoenvironment; black shale; shale gas; Southern China

1. Introduction

The development of black shales mostly occurs during particular periods of geological history, and they not only record the changes in the paleoenvironment, paleoclimate, and paleontology, but also contain large quantities of metal-rich minerals and oil and gas resources [1–4]. The Early Cambrian Niutitang black shale in the Guizhou Province and northwest Hubei Province is an iconic deposit dating from the Neoproterozoic to the Early Paleozoic Era in Southern China [5,6]. This black shale, which is rich in shale gas resources, contains 3.55×10^{12} m^3 shale gas in the Upper Yangtze Region alone [7]. Shale gas production in China reached 228×10^8 m^3 per annum in 2021, but this shale gas was mainly extracted from the Late Ordovician Wufeng and Early Silurian Longmaxi formations in the Sichuan Basin [4,8]. In contrast, the development of shale gas from the

Early Cambrian Niutitang black shale has been very limited, even though abundant gas is preserved in it.

The abundance, type, and maturity of organic matter are three key parameters determing gas potentiality in shales [9,10]. Organic matter accumulation is the first requirement for generating and preserving shale gas, and accurately identifying the organic matter accumulation mechanism is helpful for shale gas development. Previous studies have focused on the depositional environment, lithofacies, mineral composition and genesis, and pore structure of the Early Cambrian Niutitang black shale. These studies concluded that this black shale is rich in organic matter, dominated by quartz and clay minerals, and strongly heterogeneous [7,11–18]. However, the mechanism of organic matter accumulation in this black shale has been debated for many years because it is a very complicated biogeochemical process [19–21].

Currently, it is widely acknowledged that organic matter accumulation and preservation in sedimentary strata are determined by a combination of primary productivity, paleoredox conditions, and sedimentation rate in sedimentary basins [21–24]. Primary productivity determines the input quantity of the organic matter, while the preservation of this organic matter is controlled by the paleoredox conditions and sedimentation rate [22,25]. A high sedimentation rate not only promotes organic matter preservation by reducing its exposure time under oxic conditions, but also dilutes the organic matter [23]. Moreover, other factors such as hydrothermal activity, upwelling, river input, continental weathering, and water mass restrictions also influence organic matter accumulation [6,21,26]. Owing to the combination of the abovementioned influences, it is difficult to accurately explain the organic matter accumulation mechanisms of black shale.

Taking the Early Cambrian Niutitang black shale as an example, some researchers have reported that an anoxic paleoenvironment is the most significant factor affecting organic matter accumulation [27,28], while other researchers argue that a high productivity is the most important factor [14,29]. Still, others have concluded that the organic matter preserved in this black shale was accumulated by a combination of ideal preservation conditions and a high productivity [30,31]. It is worth noting that during the deposition of the Early Cambrian Niutitang black shale, the Upper Yangtze Region mainly included the intra-platform basin, slope, and deep basin settings [32], and the redox conditions, hydrothermal activity, and amount of terrigenous flux in these settings were different [6,26,33,34]. Moreover, Zhu et al. (2019) [7] reported that the total organic carbon (TOC) content of the Early Cambrian Niutitang black shale decreases from the intra-platform basin margin, through the slope and inner intra-platform basin, to the deep basin and shore settings in the Upper Yangtze Region (Figure 1). Thus, the organic matter accumulation mechanism may be completely different for the various settings. Previous studies have discussed the distribution of the organic matter content in the Early Cambrian Niutitang black shale [14,26,35–37], while the differences in organic matter accumulation and its mechanism among the contemporaneous sediments of the different sedimentary settings (such as intra-platform basin, slope, and deep basin) are still unclear.

The primary goals of this study were to determine the characteristics of the paleoredox conditions, paleoproductivity, paleoclimate, hydrothermal activity, terrigenous flux, and water mass restrictions of the different settings (intra-platform basin, through slope, to deep basin) and to compare the organic matter accumulation mechanisms among these settings during the sedimentation of the Early Cambrian Niutitang black shale. Geochemical proxies such as U/Th, V/(V + Ni), V/Cr, Ni/Co, biogenic Ba (Ba_{bio}) and Si (Si_{bio}), and δCe were used to reconstruct the paleoenvironment. Then, organic geochemical methods were combined with these geochemical proxies to investigate the organic matter accumulation mechanism. The results of this study improve our understanding of the distribution of the organic matter in the Early Cambrian Niutitang black shale and provide important information for predicting favorable areas for shale gas production.

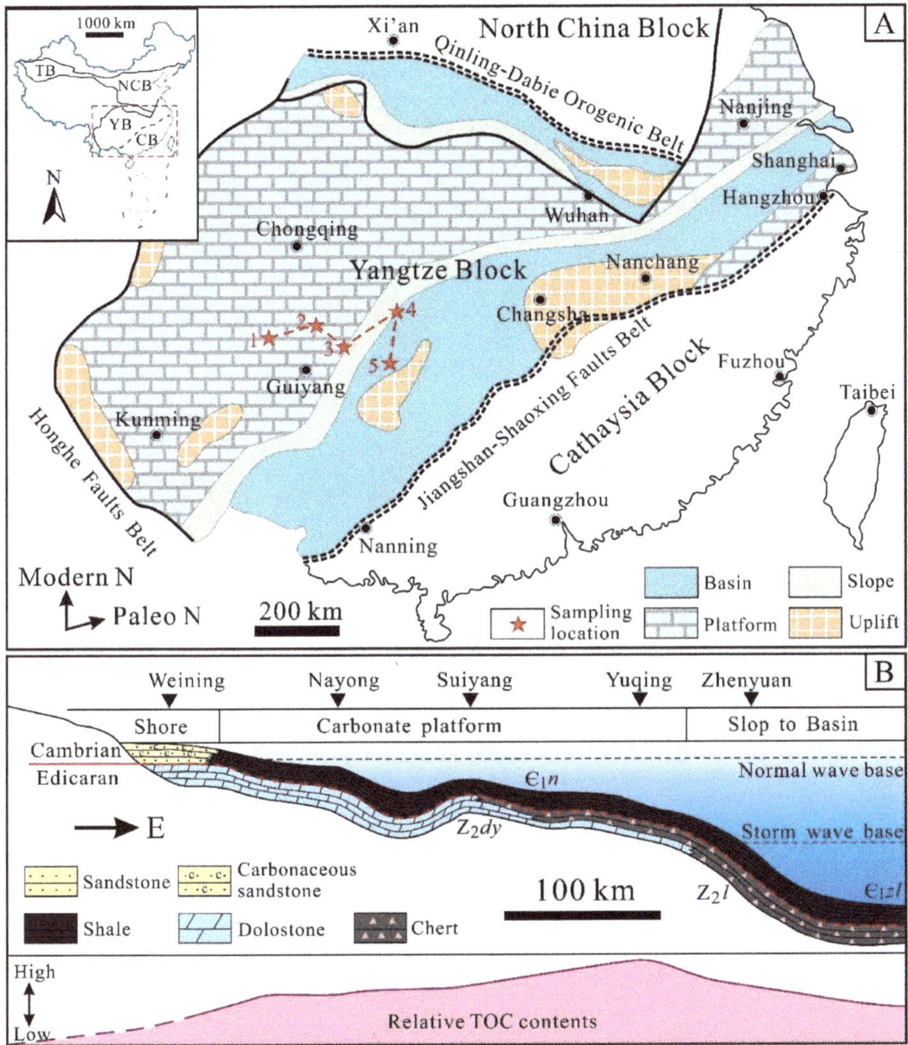

Figure 1. (**A**) Paleoenvironment reconstruction of the Yangtze Block during the Ediacaran–Cambrian transition stage illustrating the geological setting of the study area (modified from [32,38]). The red stars are the sample locations. 1—Yankong section, 2—Songlin section, 3—Yonghe section, 4—well ZK102, and 5—well ZK205. (**B**) Sketch showing the sedimentary environment and the relative TOC contents of the Lower Cambrian black shale, Guizhou Province (modified from [39], and the TOC data are from [7]). Z_2dy—the Ediacaran Dengying Formation, Z_2l—the Ediacaran Laobao Formation, $\in_1 n$—the Cambrian Niutitang Formation, and $\in_1 zl$—the Cambrian Zhalagou Formation.

2. Geological Setting

The South China Continent and the Jiangnan orogeny gradually developed around 880–860 Ma as a result of the collision between the Yangtze Block and Cathaysia Block [40,41]. This coupled South China Continent was then subjected to intense rifting during 850–820 Ma, resulting in its gradual separation into the Yangtze Block and Cathaysia Block once again; the Nanhua Basin was developed between these two blocks [41,42]. In the Ediacaran, especially during the Ediacaran–Cambrian transition stage, the Yangtze Block was subjected to intense extension and block tilting, resulting in the development of a carbonate platform on the paleohigh surrounded by a large-scale open siliceous basin (Figure 1A) [14,32,43].

The Yangtze Block developed sedimentary setting belts, generally including an intra-platform basin, slope, and deep basin, from northwest to southeast during the Ediacaran–Cambrian transition stage [32]. Interestingly, the Early Cambrian black shale has completely different stratigraphic contact relationships in the intra-platform basin compared to the slope and deep basin [6,18]. In the intra-platform basin setting, the Early Cambrian black shale of the Niutitang Formation unconformably overlies the dolostone of the Ediacaran Dengying Formation (Figures 1B and 2). However, this black shale conformably overlies the chert of the Ediacaran Laobao Formation (Figures 1B and 2)–a contemporaneous sediment of the Dengying dolostone–in the slope and deep basin settings [18,44,45]. Thus, this black shale is considered to be the key layer representing the transition from the Ediacaran to the Cambrian in the Yangtze Block and the change in the environment and climate [6,11,26,46], and it is one of the most significant shale gas reservoirs in China [4,7].

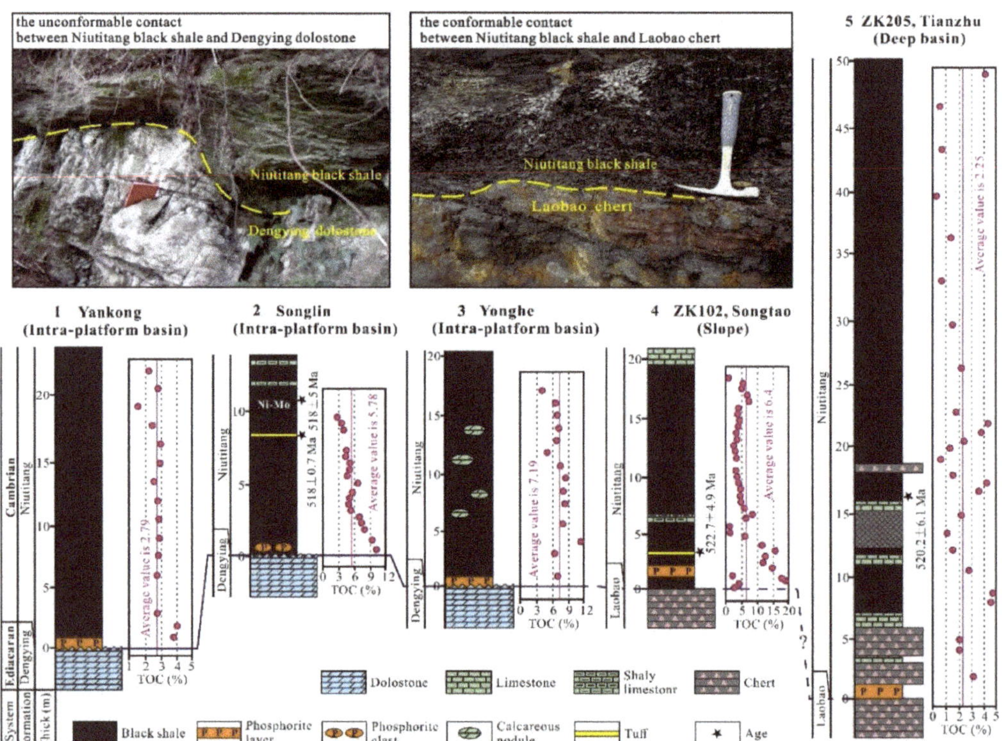

Figure 2. Contact relationship between the Ediacaran and Cambrian strata, and the stratigraphic columns and TOC contents of the Early Cambrian black shale. The ages are cited from [47–49].

3. Materials and Methods

3.1. Sample Collection and Preparation

A total of 110 fresh black shale samples were collected from three sections and two wells for the geochemical analyses conducted in this study (Figure 1A), and all the samples were sealed in plastic bags to ensure as little contamination as possible. Among them, 15 samples were from the Yankong section in northwestern Guizhou Province; 18 samples were from the Songlin section in northern Guizhou Province; 14 samples were from the Yonghe section in southern Guizhou Province; and 37 and 26 samples were from well ZK102 in northeastern Guizhou Province and well ZK205 in southeastern Guizhou

Province, respectively. All the potentially weathered surfaces, visible pyrite grains, and post-depositional veins were removed from each sample. Then, the samples were cut into small pieces. Approximately 100 g of the freshest pieces was crushed to powder (<200 mesh) using a tungsten carbide crusher.

3.2. Analytical Methods

The TOC content and element concentrations (including major, trace, and rare earth elements) of all sample powders were measured after they were acidified using 5% HCl to remove the carbonates, and neutralized to a pH of 7.0 by adding deionized water. The measurement of TOC content was conducted at the Central Laboratory of Geological Sciences, China National Petroleum Corporation (CNPC) Petroleum Exploration and Development Institute, using a LECO CS-230 sulfur-carbon analyzer.

The analysis of the element concentrations was conducted at the Institute of High Energy Physics, Chinese Academy of Science, following four Chinese National Standards (GBW07103, GBW07104, GBW07107, and GBW07108). Prior to the major element analysis, about 200 g of sample powder was dried at 100 °C to remove any moisture. A $Li_2B_4O_7$-$LiBO_2$-LiF mixture (about 6 g), ammonium nitrate (0.4 g), and 1 mg of sample powder were accurately weighted into a crucible, and the crucible was then slightly shaken to ensure that the solid particles within it were completely mixed. About 0.25 mL of lithium bromide solution was added to the mixture. Subsequently, the mixture in the crucible was fused at 1080 °C for 10 min to generate a glass bead, and the generated glass bead was measured with an X-ray fluorescence spectrometer.

For the trace and rare earth element content analyses, about 300 mg of sample powder was ashed at 850 °C for 8 h to remove any volatiles before the analysis. A mixture of HF, HNO_3, and HCl was used to dissolve each ashed sample (about 50 mg). After this digestion, the solution was diluted using 2% nitric acid and then measured using an inductively coupled plasma mass spectrometer. The procedures for the element analyses have also been described by Li et al. (2010, 2020) [11,12]. The results of the trace and rare earth element analyses have also been presented in our previous publications [15,18].

4. Results

4.1. Total Organic Carbon Content

The TOC contents of the shale samples are shown in Figure 2, and they vary from 0.29% to 28% with an average of 4.93%. Among the sampling profiles, the TOC contents decrease from the Yonghe section, through the Songlin section and well ZK102, to the Yankong section and well ZK205. In the Yonghe section, the average TOC content is as high as 7.19%. In the Songlin section and well ZK102, the TOC contents are 2.93–10.2% (5.78% on average) and 1.19–28% (6.4% on average), respectively. In the Yankong section and well ZK205, the TOC contents are as low as 1.66–4% (2.79% on average) and 0.29–4.7% (2.25% on average), respectively (Figure 2).

4.2. Major Elements

The experimental results of the major elements are presented in the Supplementary Materials. In general, SiO_2 and Al_2O_3 are the major components of the shale samples from the Yankong section (average SiO_2 and Al_2O_3 contents of 57.65% and 15.5%, respectively), Songlin section (average content of SiO_2 is 65.9% and Al_2O_3 is 12.92%), Yonghe section (average SiO_2 and Al_2O_3 contents of 72% and 8.93%, respectively), and well ZK102 (average SiO_2 and Al_2O_3 contents of 52.7% and 11.13%, respectively), while the shale samples from well ZK205 have much lower Al_2O_3 contents (0.21–8.17%, 3.75% on average).

4.3. Trace Elements

The trace element data are presented in the supplementary materials. The total trace element concentrations (40 trace elements) of the shale samples range from 1058.27 to 46,941.16 ppm. Among the sampling profiles, the Yonghe section has the highest trace element concentrations, with an average concentration of 8557.21 ppm. In addition, the shale samples from the intra-platform basin setting (average concentrations of 3223.94 and 4809.08 ppm for the Yankong section and the Songlin section, respectively) have much lower trace element concentrations than the samples from the slope and deep basin settings (average concentrations of 7459.64 and 7982.13 ppm for wells ZK102 and ZK205, respectively).

The enrichment factor is a good indicator for elemental enrichment or depletion and the enrichment factors of trace elements were calculated using the following

$$X_{ef} = (X/Sc)_{sample}/(X/Sc)_{UCC}$$

where X is the target trace element (such as Ni, Ba, Mo, U, and V), and UCC is the average composition of the upper continental crust [50]. The calculated results show that (1) the shale samples in this study are rich in Ni, Mo, Re, U, V, Cr, Cu, Zn, Ga, Cd, In, Sb, and Ba in contrast to the upper continental crust (Figure 3); and (2) the shale samples indicating the intra-platform basin setting have higher enrichment factors than the shale samples correlated to the slope and deep basin settings (Figure 3).

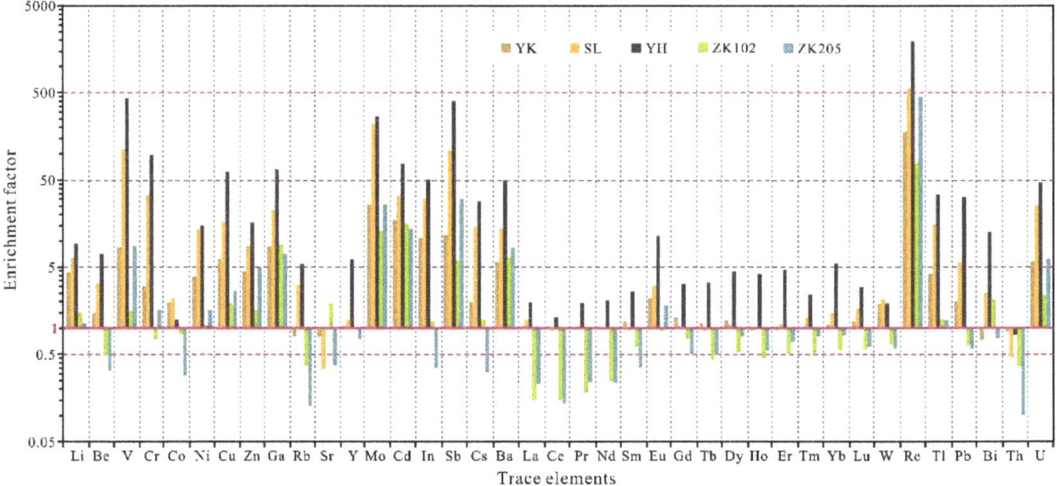

Figure 3. The average enrichment factors of trace elements in the Early Cambrian black shale.

4.4. Rare Earth Elements (REEs)

The rare earth element (REE) concentrations are presented in the supplementary materials and the REEs pattern of NASC was used to normalize the REEs data in this study [51]. Figure 4 shows the REE distribution patterns, total REE ($\sum$REE) contents, and the ratios of light REEs and heavy REEs (LREE/HREE) of the shale samples in this study. The $\sum$REE values of the shale samples vary from 2.46 to 230.02 ppm (47.17 ppm on average), and the LREE/HREE ratios are 1.87–48.87, with an average of 12.15, indicating light REE enrichment. The Yankong section has the highest average $\sum$REE value of 85.48 ppm. The average $\sum$REE values of samples from wells ZK102 and ZK205 (50.37 and 43.47 ppm, respectively) are much lower than in the Yankong section, but they are significantly higher than in the Songlin and Yonghe sections (28.68 and 28.31 ppm, respectively). The LREE/HREE ratios decrease from the intra-platform basin to the slope and deep basin. The Yankong and Songlin sections of the intra-platform basin setting have ratios of 25.07 and 23.47, respectively. In contrast, the ratios of wells ZK102 (slope setting) and ZK205 (deep basin setting) are as low as 7.07 and 5.71, respectively. The Yonghe section is located in the transition zone between the intra-platform basin and slope (Figures 1 and 2), and it has a medium ratio (9.15) among the sampling profiles studied (Figure 4).

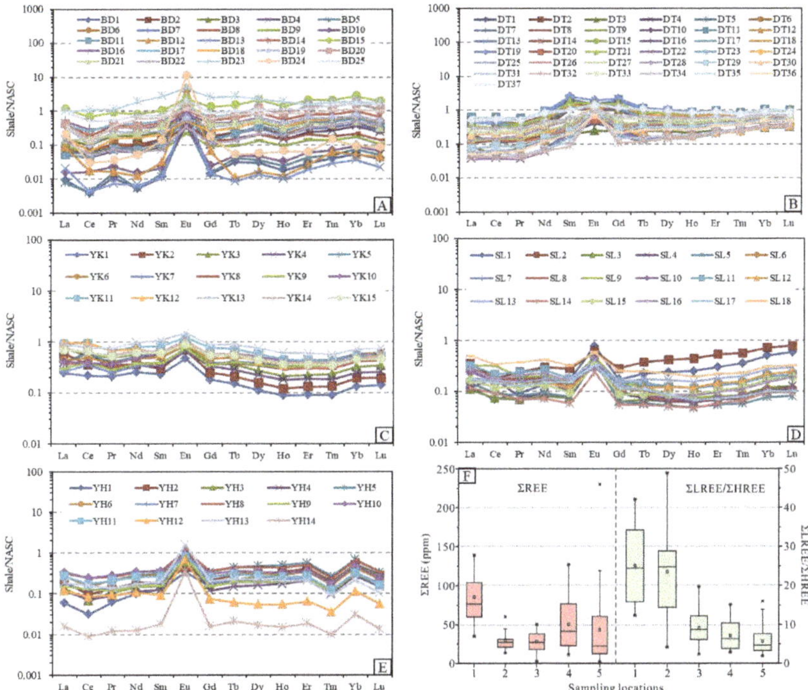

Figure 4. Rare earth element (REE) distribution patterns for the Early Cambrian black shale, and REE contents and light REE to heavy REE (LREE/HREE) ratios. (**A**) ZK205 well, (**B**) ZK102 well, (**C**) Yankong section, (**D**) Songlin section, (**E**) Yonghe section, (**F**) REE contents and light REE to heavy REE (LREE/HREE) ratios (1—Yankong section, 2—Songlin section, 3—Yonghe section, 4—well ZK102, and 5—well ZK205. Box-plot: asterisks are the maximum and minimum values, hollow square is average value).

5. Discussion

5.1. Hydrothermal Influences

Hydrothermal fluids carry many ore-forming elements (such as Mn, Ba, Cu, Pb, Fe, and Zn), and these elements can be transported and precipitated in response to changes in the physical and chemical conditions [52,53]. As a result, the contents of these elements in the sediments will be abnormally and radially distributed around the center of the hydrothermal fluid input channel. Thus, anomalous abundances of these elements are one of the most important geochemical characteristics of hydrothermal sediments and can be used to distinguish between hydrothermal sediments and normal sediments [26,54–56]. In contrast to normal sediments, hydrothermal sediments are generally enriched in elements such as Si, Mn, P, Pb, Zn, Cu, Ba, Fe, Sr, and U, while they are depleted in elements such as Al, Ti, Mg, Cr, Th, Zr, and Rb. As shown in Figure 5, the contents of the major element set Si+Fe+Mn+P and trace element set Cu+Pb+Zn+Ba+Sr+U in the shale samples are generally higher than those in the crust, while the contents of the major element set Al+Ti+Mg and trace element set Cr+Th+Rb are lower than those in the crust, indicating that the sedimentation of this shale was related to a hydrothermal event. In particular, in the slope and deep basin regions, the thick chert beds at the top of the Ediacaran strata (Figures 1B and 2) indicate that the hydrothermal activities mainly occurred in these regions during the Ediacaran–Cambrian transition stage. Wei et al. (2018) [57] suggested that the mounded cherts in the Yangtze Block were deposited in a syngenetic fault zone on the margin of the basin, and they are the results of strong submarine hydrothermal activities.

In this study, bedded cherts were observed in the slope (Laobao Formation, well ZK102) and deep basin (Laobao Formation and the bottom of the Niutitang Formation, well ZK205) regions, but no mounded chert was observed in these regions, indicating that hydrothermal activities occurred in the slope and deep basin regions during the Ediacaran–Cambrian transition stage, and the wells (ZK102 and ZK205) are not located in the central area of the hydrothermal activities.

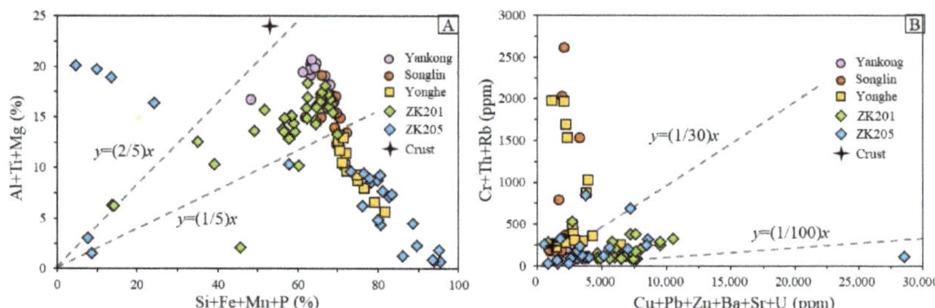

Figure 5. Relationships between different (**A**) major element sets and (**B**) trace element sets for the Early Cambrian black shale.

Sediments affected by hydrothermal activities are characterized by (1) relatively low REE abundances, (2) pronounced enrichment of HREEs relative to LREEs, and (3) positive Eu anomalies [51,53,58]. The REE distribution patterns and the ranges of the REE abundances and LREE/HREE ratios of the shale samples are shown in Figure 4. The samples from the two wells (wells ZK102 and ZK205) and the Yonghe section have relatively low REE abundances and $\sum$LREE/$\sum$HREE ratios, and large positive Eu anomalies, confirming the effects of hydrothermal activities on the Early Cambrian Niutitang black shale from the intra-platform basin margin, through the slope, to the deep basin. It should be noted that the samples from the Songlin section (intra-platform basin setting; Figures 1 and 2) also have relatively low REE abundances and positive Eu anomalies. However, these samples have high proportions of LREEs and Cr+Th+Rb concentrations (Figure 5B), suggesting that their positive Eu anomalies and low REE abundances did not originate from hydrothermal activities. The samples from the intra-platform basin setting may have been significantly affected by the terrigenous flux.

5.2. Paleoclimate and Terrigenous Flux

In general, a warm and humid tropical climate triggers stronger chemical weathering than a cold and dry climate. The chemical index of alteration (CIA) is one of the most effective proxies evaluating the intensity of chemical weathering, which is expressed as follows:

$$CIA = molar[Al_2O_3/(Al_2O_3 + CaO* + Na_2O + K_2O)] \times 100,$$

where CaO* represents the CaO in the silicate minerals only [59,60], which has been corrected using the P_2O_5 data (CaO* = mole (CaO-P_2O_5) $\times$ 10/3). If the remaining molar amount of CaO is less than that of Na_2O, CaO* should be used as the CaO value; otherwise, the CaO* should be equivalent to the Na_2O value [61]. It should be noted that K-metasomatism of shales has a large effect on the CIA value; thus, it is necessary to make corrections using the $Al_2O_3 - (CaO* + Na_2O) - K_2O$ (A-CN-K) ternary diagram [62].

The shale samples from the intra-platform basin (including the Yankong, Songlin, and Yonghe sections; Figures 1 and 2) generally exhibit moderate–high CIA values (Yankong section: 73.1 $\pm$ 2.7; Songlin section: 75.0 $\pm$ 2.7; and Yonghe section: 78.7 $\pm$ 7.0; Figure 6), suggesting that their source rocks underwent intense weathering under a warm and humid climate condition. In contrast, the slope (ZK102 well) to deep basin (ZK205 well) setting samples have low–moderate CIA values of 69.2 $\pm$ 7.1 in well ZK102 and 69.7 $\pm$ 8.4 in well ZK205 (Figure 6). These values are indicative of weak chemical weathering in the source area, likely under a cooler and drier condition [6,12]. This spatiotemporal chemical weathering intensity pattern implies that the influx of terrigenous sediments to the intra-platform basin may have been different from that to the slope and deep basin settings. Combining results from this study and previous research, we suggest that the influx to the intra-platform basin was mainly from the Yangtze Block, while that to the slope and deep basin region was partly from the Cathaysia Block [12,16,63].

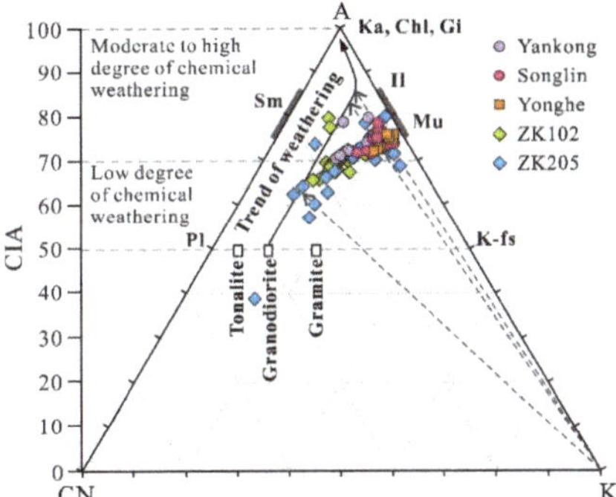

Figure 6. A-CN-K ternary diagram showing the diagenetic K-metasomatism of the Early Cambrian black shales. The associated range of the CIA values is also indicated. A—Al_2O_3; CN—CaO* + Na_2O; K—K_2O; CIA—chemical index of alteration; Ka - kaolinite; Chl—chlorite; Gi—gibbsite; Sm—smectite; Il—illite; Mu—muscovite; Pl—plagioclase; and K-fs—K-feldspar.

Elements such as Al, Ti, Th, and Zr are mainly from continental rocks and are not significantly affected by diagenesis and weathering; thus, they are considered to be indicators of the terrigenous flux [64]. Aluminum (Al) generally exists as aluminosilicate clay minerals, while Ti, Th, and Zr are usually contained in clays and heavy minerals [65]. Therefore, the Ti/Al, Th/Al, and Zr/Al ratios can indicate the input of coarse terrigenous clasts. In this study, Al, Ti, and Th were used as indicators of the terrigenous flux of the Early Cambrian Niutitang black shale, and the results are shown in Figure 7. The Yankong, Songlin, and Yonghe sections, along with well ZK102, exhibit high Al and Ti concentrations, especially in the Yankong (average concentrations of Al = 15.50%; Ti = 0.71%) and Songlin (Al = 12.92%; Ti = 0.68%) sections, which are much higher than those of the samples from well ZK205 (Al = 3.75%; Ti = 0.25%). The changes in these contents indicate that the terrigenous input to the deep basin was much lower than that to the intra-platform basin. It should be noted that Al, Ti, and Th have significant negative correlations with the TOC content for the intra-platform basin (Figure 7), suggesting that the terrigenous input had a dilution effect on the organic matter enrichment of the shale.

5.3. Sedimentation Rate

Rare earth elements (REEs) are transported into the ocean with suspended particles and detrital minerals, and the residence time of these particles and minerals in the sea water determines the degree of differentiation of the REEs [66]. Generally, a fast sedimentation rate corresponds to a moderate La_N/Yb_N ratio and a low Nd concentration, while a slow sedimentation rate results in differentiation between La and Yb and the enrichment of Nd [66–69]. As shown in Figure 8, wells ZK102 and ZK205 have lower La_N/Yb_N ratios and approximately equal Nd concentrations compared to the Yankong and Songlin sections, indicating that the sedimentation rate in the slope and deep basin settings was slightly faster than that in the intra-platform basin during the Ediacaran–Cambrian transition stage. This result conflicts with the general belief that the intra-platform basin has a faster sedimentation rate than the slope and deep basin settings. Thus, we speculate that the higher Yb concentration in shale samples from wells ZK102 and ZK205 is the result of hydrothermal activities and upwelling events.

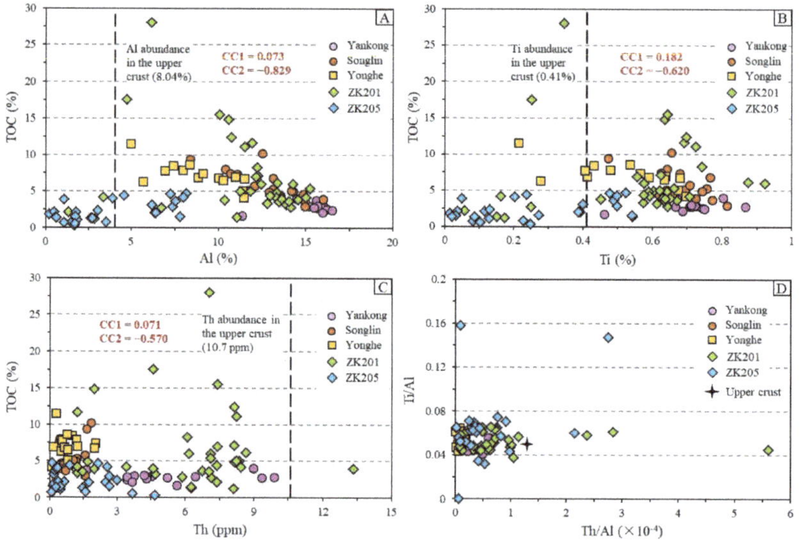

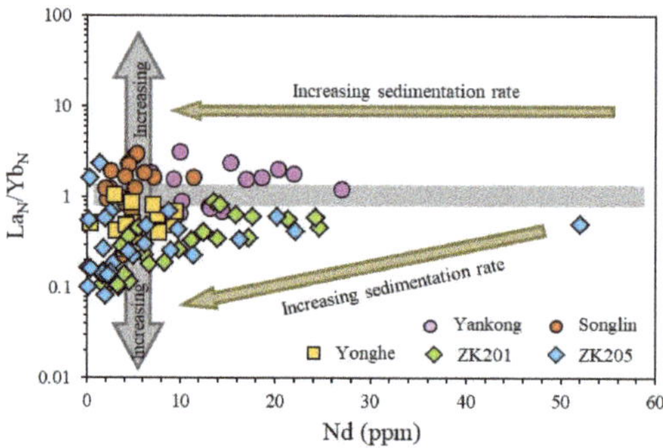

Figure 7. Plots of TOC versus (**A**) Al, (**B**) Ti, and (**C**) Th, and (**D**) plot of Ti/Al versus Th/Al. CC1—correlation coefficient of each element concentration and TOC content of all the black shales (including Yankong, Songlin, and Yonghe sections and wells ZK102 and ZK205); CC2—correlation coefficient of each element concentration and TOC content of black shales from the intra-platform basin (including Yankong, Songlin, and Yonghe sections).

5.4. Paleoproductivity

Various geochemical proxies have been effectively used to reconstruct paleoproductivity, such as the TOC content, biogenic barium (Ba_{bio}) and silica (Si_{bio}), phosphate, nutrient elements (Ni, Cu, and Zn), N and C isotopes, and biomarkers [64,70–73]. Because only a small part of the organic matter can be preserved during the stages of its passage through the water column and diagenesis, the measured TOC content of the shale is controlled by a combination of paleoproductivity, paleoredox conditions, and dilution. Some biomarkers are inaccurate when applied to over-mature shale [74]. Given the uncertainties involved in estimating the paleoproductivity based on the TOC content and biomarkers, other geochemical indicators, including Ba_{bio}, Si_{bio}, Ni+Cu+Zn, and phosphate, were used in this study (Figure 9).

Figure 8. Plot of La_N/Yb_N versus Nd showing the sedimentation rate of the Early Cambrian black shale.

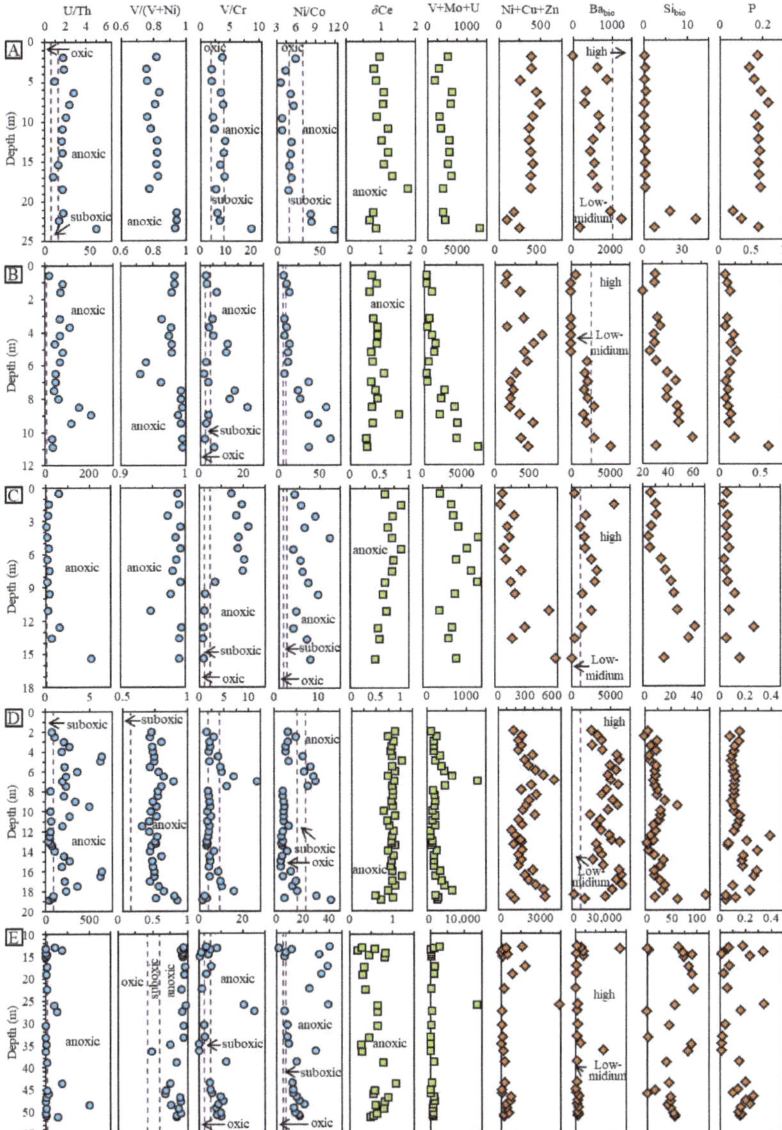

Figure 9. Stratigraphic variations in the paleoproductivity and paleoredox proxies for the (**A**) Yankong section, (**B**) Songlin section, (**C**) Yonghe section, (**D**) well ZK102, and (**E**) well ZK205. The unit for V+Mo+U, Ni+Cu+Zn, and Ba_{bio} is ppm, the unit for Si_{bio} and P is %, and U/Th, V/(V + Ni), V/Cr, Ni/Co, and δCe are dimensionless.

In a shale sample, the fraction of the Ba_{bio} is in excess relative to normal terrigenous material, and it can be calculated using the following formula [75]:

$$Ba_{bio} = Ba_{total} - \left[Al_{total} \times (Ba/Al)_{terrigenous} \right],$$

where Ba_{total} is the total amount of barium in the sample, ppm; Al_{total} is the total amount of aluminum in the sample; and $(Ba/Al)_{terrigenous}$ is the average Ba/Al ratio of the upper crust.

The Ba_{bio} value appears to originate from the decayed organic matter. This organic matter was produced by the surface productivity. In the ideal case, for shale, a high Ba_{bio} content indicates a high paleoproductivity [75]. However, barium is an unstable element under anoxic to euxinic conditions, which can be recycled into the overlying water column under the effects of sulfate reducing bacteria (such as bacterial sulfate reduction, BSR). Among the shale samples in this study, the Ba_{bio} contents of the intra-platform basin range from 11.05 to 5508.97 ppm, with an average of 1141.38 ppm, while those of the slope and deep basin range from 263.9 to 44,655.79 ppm, with an average of 4769.18 ppm (Figure 9). These results imply that (1) the slope and deep basin had a much higher primary productivity than the intra-platform basin during the Ediacaran–Cambrian transition stage in the study area, or (2) during this stage, the intra-platform basin was in anoxic condition.

Previous studies have shown that the silica content is positively correlated with the TOC content for the Early Cambrian Niutitang black shale, indicating the existence of Si_{bio} in this shale [17,76,77]. The fraction of the Si_{bio} content in the shale samples can be estimated as follows [78]:

$$Si_{bio} = Si_{total} - [Al_{total} \times (Si/Al)_{terrigenous}],$$

where Si_{total} is the total amount of silica in the sample, ppm; Al_{total} is the total amount of aluminum in the sample; and $(Si/Al)_{terrigenous}$ is the average Si/Al ratio of the upper crust.

The Si_{bio} contents of the shale samples analyzed in this study range from 0.17% to 94.44% (Figure 9). It should be noted that the samples of the slope and deep basin regions have much higher Si_{bio} contents (average of 27.52%) than those of the intra-platform basin (average of 18.08%), which indicates that the slope and deep basin environments had a higher productivity than the intra-platform basin. This conclusion is further verified by the Ni+Cu+Zn contents. As shown in Figure 9A–C, the Ni+Cu+Zn concentrations of the shale samples from the intra-platform basin region (85.11–881.34 ppm, average of 324.84 ppm) are lower than those from the slope and deep basin regions (36.97–4415.29 ppm, average of 415.55 ppm) (Figure 9D,E).

Phosphorous (P) is one of the essential nutrient elements for the growth of marine organisms on both short and long timescales, and its concentration in sediments has been used as an indicator of paleoproductivity [79,80]. In the studied shale samples, the P contents are 0.015–0.77% (Figures 9 and 10), which are higher than the content in the crust (700 ppm). These results suggest that the Upper Yangtze Region was a productive and dynamic region during the deposition stage of the Early Cambrian Niutitang black shale. As shown in Figure 10, there is a moderate correlation between P and TOC (R^2 = 0.32, n = 109), and this phenomenon probably indicates that during the deposition of this black shale, the primary productivity was high, which controlled the organic matter production.

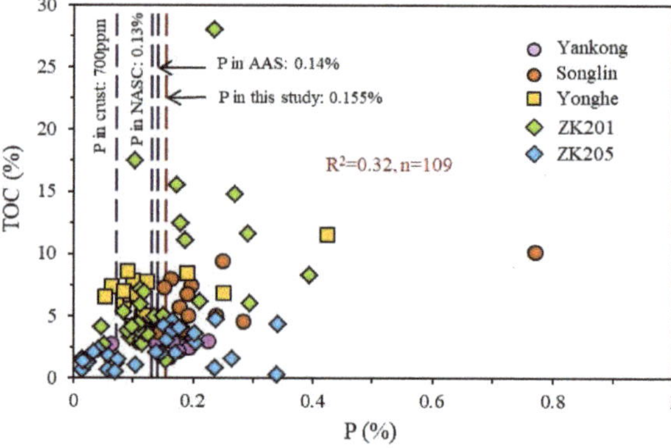

Figure 10. Plot of TOC versus P. The blue and red dotted lines indicate the average P contents of the different materials. NASC—North American Shale Composite; AAS—Average Archean Shale.

5.5. Paleoredox Conditions

The preservation of organic matter is mainly controlled by the redox conditions. In general, anoxic and euxinic conditions are beneficial for the preservation of organic matter [81–83]. The redox conditions during sedimentation can be indicated by single redox-sensitive elements (such as V, Mo, Ni, Re, U, Ce, and Eu) and the ratios of two different elements (mainly including U/Th, U/Mo, Re/Mo, V/(V + Ni), V/Cr, V/Sc, and Ni/Co). However, no single proxy can conclusively indicate the redox conditions because the distribution of these elements is affected by various factors [64,78,84,85]. In this study, indicators including U/Th, V/(V + Ni), V/Cr, and Ni/Co ratios and the V+Mo+U and δCe values, were adopted to analyze the paleo-sedimentary conditions of the Early Cambrian Niutitang black shale. The δCe value can be calculated using the following formula:

$$\delta Ce = 3 \times Ce_{sample}/(2 \times La_{sample} + Nd_{sample})$$

where Ce_{sample}, La_{sample}, and Nd_{sample} are the Ce, La, and Nd concentrations, respectively, of the shale samples [86].

Uranium (U) and V are redox-sensitive elements, and they are generally enriched under anoxic and euxinic conditions [64]. Cr and Th, unlike U and V, are not redox-sensitive elements and remain insoluble in the marine environment [78]. Therefore, the U/Th and V/Cr ratios can indicate oceanic redox conditions. Generally, U/Th > 1.25 represents anoxic conditions, and ratios as low as 0.75 represent strongly oxidizing conditions [87]. For V/Cr, >4.25 indicates anoxic conditions, 4.25–2 indicates suboxic conditions, and <2 indicates oxic conditions [81]. The V/(V + Ni) and Ni/Co ratios can also indicate redox conditions. It is widely accepted that both V/(V + Ni) and Ni/Co increase from oxic [V/(V + Ni) < 0.45; Ni/Co < 5], through suboxic [0.45 < V/(V + Ni) < 0.6; 5 < Ni/Co < 7], to anoxic [0.6 < V/(V + Ni); 7 < Ni/Co] conditions [64,78,81,88,89]. In addition, δCe > −0.1 and <−0.1 indicate anoxic and oxic conditions, respectively [90]. Since V, Mo, and U are all enriched in sediments deposited under anoxic to euxinic conditions but are dissolved and transported by oceanic water under oxic conditions, the V+Mo+U content can be used to effectively evaluate the evolution of the paleoredox conditions.

As shown in Figure 9, the U/Th ratios of the shale samples from the intra-platform basin vary from 0.88 to 209.03 (Figure 9A–C), with an average of 17.54. In contrast, the shale samples from the slope and deep basin regions have slightly lower U/Th ratios of 0.35–129.85 (Figure 9D,E) (excluding an abnormal value of 509.24 from well ZK205), with an average of 8.59. These results imply that (1) the studied area was dominated by suboxic to anoxic environments and there may have been short oxygenation events in the slope region, and (2) the intra-platform basin experienced a much stronger reducibility than the slope during the sedimentation of the Early Cambrian Niutitag black shale. These two deductions are also supported by the V/(V + Ni) ratios. The V/(V + Ni) ratios of the intra-platform basin shale samples are 0.756–0.994 (average of 0.906), indicating anoxic or even euxinic conditions of sedimentation, while those of the slope and deep basin shale samples range from 0.483 to 0.973, with an average of 0.803, indicating deposition under suboxic to anoxic conditions. This is consistent with the conclusions of our recent work, which is based on the concentrations and enrichment factors of U, Mo, Ni, and V [18].

The Early Cambrian Niutitang black shale samples have δCe values of 0.19–1.78, which are much higher than −0.1, supporting the conclusion that this shale was deposited under anoxic conditions. However, the other indicators, including V/Cr (1.54–22.24 for intra-platform basin region, and 0.13–25.22 for the slope and deep basin regions) and Ni/Co (3.65–62.98 for the intra-platform basin, and 1.65–40.63 for slope and deep basin regions), indicate that the sedimentary environment of this shale fluctuated between oxic, suboxic, and anoxic (or even euxinic) (Figure 9). Previous studies constructed a stratified and fluctuating model for the oceanic water mass on the Yangtze Block during the Ediacaran–Cambrian transition stage, arguing that several oxygenation events occurred in the background of a generally anoxic environment [11,13,34].

Taking into account the numerous indicators used in this study, including U/Th, V/(V + Ni), V/Cr, Ni/Co, V+Mo+U, and δCe, we suggest that the Early Cambrian black shale in the Upper Yangtze Region was deposited under primarily anoxic conditions, with several short oxygenation events. In contrast to the slope setting, the water environment in the intra-platform basin setting was more anoxic. The reasons for this may include (1) that hydrothermal activities and upwelling events improved the oxygen concentration of the water in the slopesetting, and (2) the intra-platform basin was relatively restricted during the sedimentation of the Early Cambrian Niutitang black shale. In addition, the typomorphic characteristics of the pyrite are significantly different in the intra-platform basin and slope to deep basin regions. The pyrite is mainly framboids in the intra-platform basin, but there is a large amount of colloidal pyrite in the slope to deep basin (Figure 11), implying a euxinic condition in the intra-platform basin [91,92].

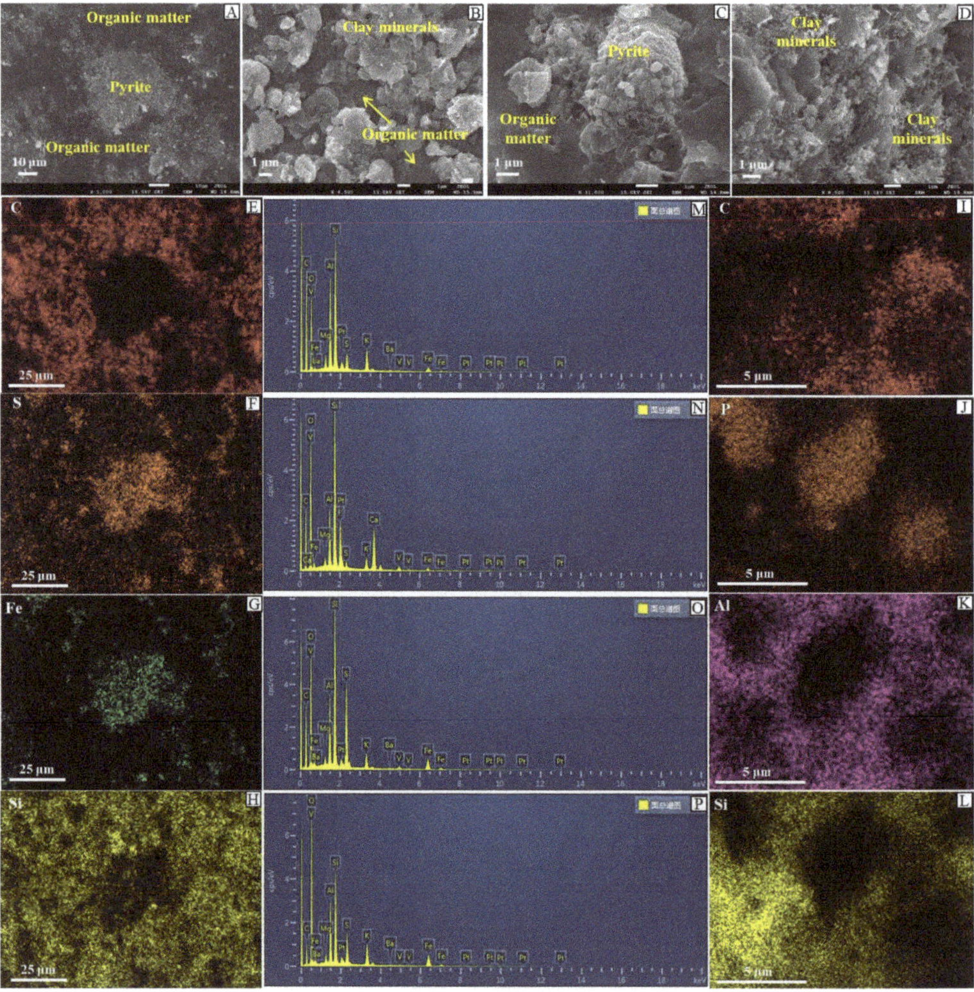

Figure 11. SEM images of the Early Cambrian black shale. (**A,B**) Samples from well ZK102; (**C,D**) sample from the Songlin section; (**E–H**) distributions of C, S, Fe, and Si corresponding to subgraph A; (**I–L**) distributions of C, P, Al, and Si corresponding to subgraph D; (**M–P**) energy spectra of subgraphs (**A–D**), respectively.

5.6. Water Mass Restriction

The degree of water mass restriction affects the sedimentary environment and the geochemical cycle, and it is a significant feature of marine systems [85]. The trace elements U and Mo can be combined to evaluate water mass restriction because (1) both of these elements have long retention times in seawater, and thus exhibit global seawater concentrations; (2) both of these elements exist in high valence states (U^{+6} and Mo^{+6}) under oxic conditions and are enriched in sediments in low valence state forms (U^{+4} and Mo^{+4}) under anoxic conditions; (3) the sediments absorb U earlier than Mo in unrestricted water masses; and (4) the absorption rate of Mo by sediments is faster than that of U in weakly and strongly restricted water masses [93–95]. Thus, a plot of U_{ef} versus Mo_{ef} can be used to evaluate the degree of water mass restriction. In addition, the supply of Mo to the basin decreases with increasing water mass restriction, resulting in low Mo/TOC ratios in a strongly restricted basin [96]. In this study, the U_{ef}-Mo_{ef} and Mo-TOC diagrams were combined to evaluate the degree of water mass restriction during the deposition of the Early Cambrian black shale in the Upper Yangtze Region.

As shown in Figure 12A, the Mo/TOC ratios of the shale samples from the intra-platform basin range from 0.37×10^{-4} to 29.74×10^{-4}, while the ratios of the slope and deep basin shale samples vary from 0.19×10^{-4} to 94×10^{-4}, suggesting that (1) the intra-platform basin was more restricted than the deep basin, and (2) the water mass restriction of the slope and deep basin regions changed more frequently than that of the intra-platform basin during the deposition of the Early Cambrian Niutitang black shale.

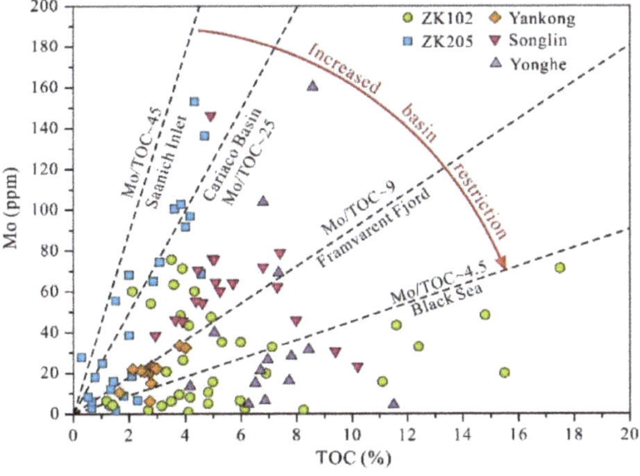

Figure 12. Plots of TOC versus Mo (modified from [94]).

5.7. Controls on Organic Matter Accumulation

The Yangtze Block experienced intense extension and block tilting, leading to the development of an intra-platform basin on the paleohigh surrounded by a large-scale open siliceous deep basin. As a result, during the terminal Ediacaran–earliest Cambrian stage of the Yangtze Block, the intra-platform basin, slope, and deep basin sedimentary setting belts were distributed from northwest to southeast (Figure 1) [14,32,47]. During this stage, upwelling events brought a large amount of nutrients to the shallow water regions, resulting in the phytoplankton in the shallow water flourishing [14,26]. In addition, organisms such as worms, sponges, and bacteria exploded around the hydrothermal centers (especially the large faults in the slope region) [26,53]. As a result, the intra-platform basin margins and slope had significantly higher primary productivities than the inner intra-platform basin (Figure 9).

Based on their profiles, the Yonghe section and ZK102 and ZK205 wells are all distributed in areas with relatively high primary productivities (see Section 5.4), and the

Energies **2022**, 15, 4551

organic matter content decreases from the Yonghe section, through well ZK102, to well ZK205. These results suggest that the organic matter accumulation was mainly controlled by the paleoproductivity in areas of high primary productivity. In these areas, the high paleoproductivity generated a large amount of organic matter, part of which was oxidized and the rest of which was diluted (Figure 13). The combined effects of the high primary productivity, sedimentation rate, and suboxic-anoxic conditions are supported by the results of the correlation analysis of the TOC content and various geochemical proxies. Taking well ZK102 as an example, none of the proxies have a correlation coefficient greater than 0.26 with the TOC content (Figure 14).

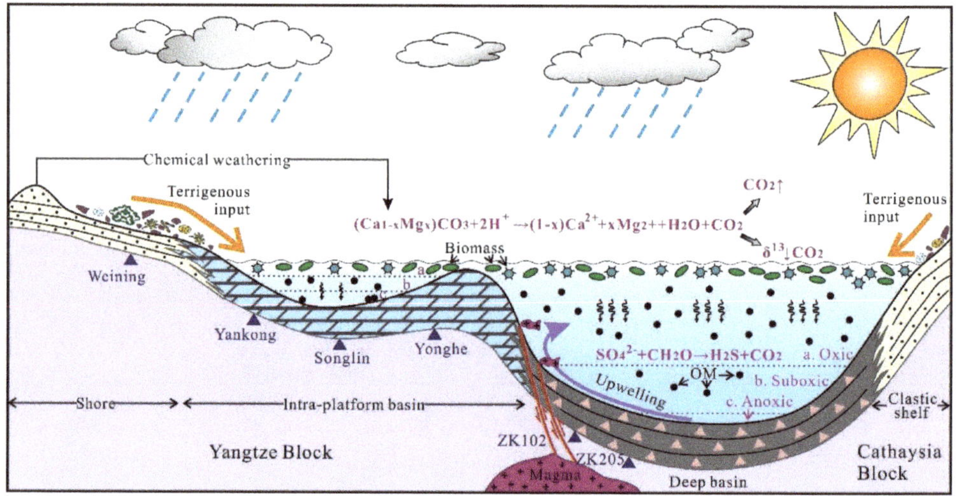

Figure 13. Schematic sketch illustrating the organic matter accumulation mechanism in the Upper Yangtze region during the deposition of the Early Cambrian black shale.

Yankong section:

	U/Th	V/(V+Ni)	V/Cr	Ni/Co	δCe	δEu	V+Mo+U	Babio	P	Si bio	Ni+Cu+Zn	ΣREE	La/Lu	Sr/Ba	Y/Ho	Al/Ti	TOC
U/Th	1.00	0.38	0.78	0.68	−0.17	0.27	0.82	−0.46	0.19	0.13	0.00	−0.24	0.24	−0.29	−0.74	0.11	0.44
V/(V+Ni)		1.00	0.57	0.89	−0.40	0.58	0.55	0.19	−0.52	0.80	−0.69	−0.34	0.65	−0.38	−0.12	−0.65	0.64
V/Cr			1.00	0.80	−0.06	0.33	0.98	−0.50	0.24	0.18	−0.12	−0.43	0.10	−0.30	−0.72	0.14	0.53
Ni/Co				1.00	−0.35	0.57	0.78	−0.07	−0.33	0.69	−0.58	−0.41	0.59	−0.34	−0.45	−0.34	0.69
δCe					1.00	−0.13	−0.02	−0.26	0.57	−0.57	0.50	−0.08	−0.45	−0.19	0.00	0.32	−0.21
δEu						1.00	0.30	0.45	−0.36	0.71	−0.66	−0.77	0.39	−0.38	−0.13	−0.52	0.80
V+Mo+U							1.00	−0.53	0.28	0.14	−0.04	−0.38	0.12	−0.38	−0.70	0.12	0.55
Babio								1.00	−0.69	0.57	−0.66	−0.09	0.33	0.23	0.55	−0.59	0.19
P									1.00	−0.75	0.80	0.14	−0.55	−0.01	−0.42	0.68	−0.19
Si bio										1.00	−0.88	−0.33	0.76	−0.17	0.07	−0.76	0.65
Ni+Cu+Zn											1.00	0.33	−0.59	−0.16	−0.04	0.53	−0.39
ΣREE												1.00	0.18	0.50	0.18	0.24	−0.60
La/Lu													1.00	−0.12	0.01	−0.54	0.44
Sr/Ba														1.00	−0.04	0.48	−0.59
Y/Ho															1.00	−0.46	−0.19
Al/Ti																1.00	−0.51
TOC																	1.00

ZK102 well:

	U/Th	V/(V+Ni)	V/Cr	Ni/Co	δCe	δEu	V+Mo+U	Babio	P	Si bio	Ni+Cu+Zn	ΣREE	La/Lu	Sr/Ba	Y/Ho	Al/Ti	TOC
U/Th	1.00	−0.06	0.21	0.02	0.51	0.80	0.17	0.54	−0.01	−0.04	0.35	−0.52	−0.54	−0.46	−0.38	0.01	0.26
V/(V+Ni)		1.00	0.35	0.67	−0.33	0.00	0.50	−0.17	−0.15	0.54	0.25	−0.24	−0.11	−0.21	−0.11	0.20	0.05
V/Cr			1.00	0.49	0.13	0.19	0.96	0.38	0.11	−0.15	0.77	−0.26	−0.41	−0.34	−0.41	0.04	0.20
Ni/Co				1.00	−0.42	0.18	0.60	−0.13	0.05	0.37	0.28	−0.37	−0.35	−0.29	−0.39	0.47	−0.16
δCe					1.00	0.37	0.03	0.58	0.04	−0.59	0.24	−0.02	−0.03	0.04	−0.27	−0.17	0.08
δEu						1.00	0.18	0.51	−0.11	−0.14	0.34	−0.64	−0.58	−0.44	−0.55	0.27	−0.10
V+Mo+U							1.00	0.28	0.08	−0.01	0.79	−0.33	−0.45	−0.40	−0.37	0.14	0.14
Babio								1.00	0.17	−0.37	0.46	−0.23	−0.33	−0.35	−0.20	−0.13	0.17
P									1.00	0.03	0.04	0.01	−0.05	−0.09	0.13	0.13	−0.13
Si bio										1.00	−0.13	−0.25	−0.15	−0.19	0.27	0.25	0.08
Ni+Cu+Zn											1.00	−0.46	−0.56	−0.59	−0.33	0.16	0.17
ΣREE												1.00	0.81	0.74	0.38	−0.40	−0.04
La/Lu													1.00	0.81	0.44	−0.32	−0.10
Sr/Ba														1.00	0.35	−0.26	−0.18
Y/Ho															1.00	−0.39	0.26
Al/Ti																1.00	−0.49
TOC																	1.00

Figure 14. Correlation coefficients of TOC content with various geochemical proxies (Purple mean positive correlation, and blue mean negative correlation).

As was discussed in Section 5.4, the inner part of the intra-platform basin (Yankong and Songlin sections) had a lower primary productivity than the slope (well ZK102), deep basin (well ZK205), and the margins of the intra-platform basin (Yonghe section) (Figure 9), indicating that the amount of initial organic matter on the inner intra-platform basin was low during the deposition of the Early Cambrian Niutitang black shale. Thus, the anoxic conditions of the marine water were necessary for the accumulation and preservation of the organic matter. As shown in Figure 2, the Songlin section has a much higher TOC content than the Yankong section, which may be because the former section experienced more anoxic conditions than the latter during the deposition of the Early Cambrian Niutitang black shale (Figures 9 and 11). In addition, the TOC contents of the Yankong section are positively correlated with the geochemical redox proxies (Figure 14), which indicates that the organic matter accumulation and enrichment in the intra-platform basin setting were controlled by the preservation conditions (Figures 9 and 13). In this study, we could not definitively conclude whether a preservation model or a productivity model was best for the Early Cambrian Niutitang black shale in the Upper Yangtze Region based on only five profiles. However, the distributions of the geochemical proxies and the TOC contents indicate that the mechanisms of the organic matter accumulation were different in the intra-platform basin setting compared to the slope to deep basin settings.

6. Conclusions

(1) For the Early Cambrian Niutitang black shale in the Upper Yangtze Region, geochemical proxies were used to compare the paleoenvironments and the mechanisms of organic matter accumulation in this shale in the intra-platform basin, slope, and deep basin settings. In contrast to the slope and deep basin settings, the intra-platform setting had a significantly higher terrigenous input and was deposited under more restricted conditions. Hydrothermal activities and upwelling events supplied a large amount of nutrients and metal elements to the surface layer of the sediments during the sedimentation of the Early Cambrian Niutitang black shale, especially in the slope and intra-platform basin margin settings.

(2) The redox proxies, mainly including U/Th, V/(V + Ni), V/Cr, V+Mo+U, and δCe, indicate that the black shale studied was deposited under suboxic-anoxic conditions. The intra-platform basin was more anoxic than the slope setting. The primary productivity was higher in the slope and deep basin settings than in the intra-platform basin. For the intra-platform basin setting, its inner part had a low productivity, but the productivity was high on its margins.

(3) In the slope and deep basin settings, the high paleoproductivity generated a large amount of organic matter, and its preservation was affected by the redox conditions. The redox conditions were the most significant factor controlling the preservation of the organic matter in the inner intra-platform basin because the paleoproductivity was lower than that in the intra-platform basin margins, slope, and deep basin settings. The intra-platform basin margins were the most favorable areas for preserving organic matter because these areas had a high paleoproductivity and anoxic conditions.

Supplementary Materials: The following supporting information can be downloaded at: https://www.mdpi.com/article/10.3390/en15134551/s1.

Author Contributions: P.X., data curation, formal analysis, and writing of the original draft. F.H., conceptualization and supervision. J.T., supervision and writing—review & editing. W.Z., sample collection and experiments. Y.F., methodology and supervision. C.G., writing—review & editing. Z.Y., writing—review & editing. K.L., writing—review & editing. K.W., writing—review & editing. All authors have read and agreed to the published version of the manuscript.

Funding: This work was financially supported by the National Natural Science Fund of China (Grant Nos. 42002166, 42162016, and 41690134) and the Guizhou Provincial Fund Project (Grant Nos. [2020]1Y161, [2020]1Y166, ZK[2021]199, ZK[2022]YIBAN106, and ZK[2022]YIBAN213).

Institutional Review Board Statement: Not applicable.

Informed Consent Statement: Not applicable.

Data Availability Statement: Available from corresponding author.

Conflicts of Interest: The authors declare no conflict of interest.

References

1. Fan, D.; Zhang, T.; Ye, J. *Black Rock Series and Related Deposits in China*; Science Press: Beijing, China, 2004.
2. Wang, Z.W.; Fu, X.G.; Feng, X.L.; Song, C.Y.; Wang, D.; Chen, W.B.; Zeng, S.Q. Geochemical features of the black shales from the Wuyu Basin, southern Tibet: Implications for palaeoenvironment and palaeoclimate. *Geol. J.* **2017**, *52*, 282–297. [CrossRef]
3. Shi, C.H.; Cao, J.; Han, S.C.; Hu, K.; Bian, L.Z.; Yao, S.P. A review of polymetallic mineralization in lower Cambrian black shales in South China: Combined effects of seawater, hydrothermal fluids, and biological activity. *Palaeogeogr. Palaeoclimatol. Palaeoecol.* **2021**, *561*, 110073. [CrossRef]
4. Zou, C.N.; Zhao, Q.; Cong, L.Z.; Wang, H.Y.; Shi, Z.S.; Wu, J.; Pan, S.Q. Development progress, potential and prospect of shale gas in China. *Nat. Gas Ind.* **2021**, *41*, 1–14, (In Chinese with English abstract).
5. Guizhou Geological Survey. In *Regional Geology of China*; Geology Press: Beijing, China, 2017.
6. Fu, Y.; Xia, P.; Long, Z.; Zhang, G.J.; Qiao, W.L.; Guo, C.; Yang, Z. Continental weathering of Yangtze area during Ediacaran (Sinian)–Cambrian transition stage: Advances and prospects. *Geol. Rev.* **2021**, *67*, 1–18, (In Chinese with English abstract).
7. Zhu, L.J.; Zhang, D.W.; Zhang, J.C.; Wang, G.L.; Yang, T.B.; Chen, H.G. *Geological Theroy and Practice of Shale Gas in Paleozoic Passive Continental Margin, Eastern of Upper Yangtze*; Science Press: Beijing, China, 2019.
8. Zou, C.N.; Yang, Z.; Dong, D.Z.; Zhao, Q.; Chen, Z.H.; Feng, Y.L.; Li, J.R.; Wang, X.N. Formation, Distribution and Prospect of Unconventional Hydrocarbons in Source Rock Strata in China. *Earth Sci.* **2022**, *47*, 1517–1533, (In Chinese with English abstract).
9. Hao, F.; Zou, H.Y. Cause of shale gas geochemical anomalies and mechanisms for gas enrichment and depletion in high-maturity shales. *Mar. Pet. Geol.* **2013**, *44*, 1–12. [CrossRef]
10. Zhang, K.; Song, Y.; Jia, C.Z.; Jiang, Z.X.; Han, F.L.; Wang, P.F.; Yuan, X.J.; Yang, Y.M.; Zeng, Y.; Li, Y.; et al. Formation mechanism of the sealing capacity of the roof and floor strata of marine organic-rich shale and shale itself, and its influence on the characteristics of shale gas and organic matter pore development. *Mar. Pet. Geol.* **2022**, *140*, 105647. [CrossRef]
11. Li, C.; Love, G.D.; Lyons, T.W.; Fike, D.A.; Sessions, A.L.; Chu, X.L. A Stratified Redox Model for the Ediacaran Ocean. *Science* **2010**, *328*, 80–83. [CrossRef]
12. Li, C.; Zhang, Z.H.; Jin, C.S.; Cheng, M.; Wang, H.Y.; Huang, J.H.; Algeo, T.J. Spatiotemporal evolution and causes of marine euxinia in the early Cambrian Nanhua Basin (South China). *Palaeogeogr. Palaeoclimatol. Palaeoecol.* **2020**, *546*, 109676. [CrossRef]
13. Jin, C.S.; Li, C.; Algeo, T.J.; Planavsky, N.J.; Cui, H.; Yang, X.L.; Zhao, Y.L.; Zhang, X.L.; Xie, S.C. A highly redox–heterogeneous ocean in South China during the early Cambrian (∼529–514 Ma): Implications for biota–environment co–evolution. *Earth Planet. Sci. Lett.* **2016**, *441*, 38–51. [CrossRef]
14. Yeasmin, R.; Chen, D.Z.; Fu, Y.; Wang, J.G.; Guo, Z.H.; Guo, C. Climatic–oceanic forcing on the organic accumulation across the shelf during the Early Cambrian (Age 2 through 3) in the mid–upper Yangtze Block, NE Guizhou, South China. *J. Asian Earth Sci.* **2017**, *134*, 365–386. [CrossRef]
15. Zhou, W.X. *Sedimentary Environment and Geochemical Characteristics of the Lower Cambrian Black Rock Series in Northern Guizhou*; Guizhou University: Guiyang, China, 2017.
16. Zhai, L.N.; Wu, C.D.; Ye, Y.T.; Zhang, S.C.; Wang, Y.Z. Fluctuations in chemical weathering on the Yangtze Block during the Ediacaran–Cambrian transition: Implications for paleoclimatic conditions and the marine carbon cycle. *Palaeogeogr. Palaeoclimatol. Palaeoecol.* **2018**, *490*, 280–292. [CrossRef]
17. Xia, P.; Li, H.N.; Fu, Y.; Qiao, W.L.; Guo, C.; Yang, Z.; Huang, J.Q.; Mou, Y.L. Effect of lithofacies on pore structure of the Cambrian organic–rich shale in northern Guizhou, China. *Geol. J.* **2021**, *56*, 1130–1142. [CrossRef]
18. Fu, Y.; Zhou, W.X.; Wang, H.J.; Qiao, W.L.; Ye, Y.T.; Jiang, R.; Wang, X.M.; Su, J.; Li, D.; Xia, P. The relationship between environment and geochemical characteristics of black rock series of Lower Cambrian in northern Guizhou. *Acta Geol. Sin.* **2021**, *95*, 536–5483, (In Chinese with English abstract).
19. Kuypers, M.M.M.; Pancost, R.D.; Nijenhuis, I.A.; Sinninghe Damste, J.S. Enhanced productivity led to increased organic carbon burial in the euxinic North Atlantic basin during the late Cenomanian oceanic anoxic event. *Paleoceanography* **2002**, *17*, 1051. [CrossRef]
20. Hartnett, H.E.; Devol, A.H. Role of a strong oxygen–deficient zone in the preservation and degradation of organic matter: A carbon budget for the continental margins of northwest Mexico and Washington State. *Geochim. Cosmochim. Acta* **2003**, *67*, 247–264. [CrossRef]
21. Ding, X.J.; Liu, G.D.; Huang, Z.L.; Lu, X.J.; Chen, Z.L.; Gu, F. Controlling function of organic matter supply and preservation on formation of source rock. *Earth Sci.* **2016**, *41*, 832–842. (In Chinese with English abstract)
22. Pedersen, T.F.; Calvert, S.E. Anoxia vs. Productivity: What Controls the Formation of Organic–Carbon–Rich Sediments and Sedimentary Rocks? *AAPG Bull.* **1990**, *74*, 454–466.
23. Tyson, R.V. Sedimentation rate, dilution, preservation and total organic carbon: Some results of a modelling study. *Org. Geochem.* **2001**, *32*, 333–339. [CrossRef]
24. Tyson, R.V.; Harris, N.B. *The "Productivity Versus Preservation" Controversy: Cause, Flaws, and Resolution, The Deposition of Organic–Carbon–Rich Sediments: Models, Mechanisms, and Consequences*; SEPM Special Publication: Tulsa, OK, USA, 2005.

25. Lu, Y.B.; Jiang, S.; Lu, Y.C.; Xu, S.; Shu, Y.; Wang, Y.X. Productivity or preservation? The factors controlling the organic matter accumulation in the late Katian through Hirnantian Wufeng organic–rich shale, South China. *Mar. Pet. Geol.* **2019**, *109*, 22–35. [CrossRef]

26. Xia, P.; Fu, Y.; Yang, Z.; Guo, C.; Huang, J.Q.; Huang, M.Y. The relationship between sedimentary environment and organic matter accumulation in the Niutitang black shale in Zhenyuan, northern Guizhou. *Acta Geol. Sin.* **2020**, *94*, 947–956. (In Chinese with English abstract)

27. Algeo, T.J.; Maynard, J.B. Trace-element behavior and redox facies in core shales of Upper Pennsylvanian Kansas-type cyclothems. *Chem. Geol.* **2004**, *206*, 289–318. [CrossRef]

28. Li, J. *Study on Paleo-Environmental Reconstruction and Organic Matter Accumulation of the Lower Cambrian Niutitang Formation in Northern Guizhou*; China University of Geosciences: Beijing, China, 2018.

29. Wei, H.R.; Yang, R.D.; Gao, J.B.; Chen, J.Y.; Liu, K.; Cheng, W. New Evidence for Hydrothermal Sedimentary Genesis of the Ni–Mo Deposits in Black Rock Series of the Basal Cambrian, Guizhou Province: Discovery of Coarse–Grained Limestones and its Geochemical Characteristics. *Acta Geol. Sin.* **2012**, *86*, 51–61.

30. Liu, Z.H.; Zhuang, X.G.; Teng, G.E.; Xie, X.M.; Yin, L.M.; Bian, L.Z.; Feng, Q.L.; Algeo, T.J. The lower cambrian niutitang formation at yangtiao (guizhou, sw china): Organic matter enrichment, source rock potential, and hydrothermal influences. *J. Pet. Geol.* **2015**, *38*, 411–432. [CrossRef]

31. Wang, N.; Li, M.J.; Tian, X.W.; Hong, H.T.; Liu, P.; Chen, G.; Wang, M.L. Main factors controlling the organic matter enrichment in the Lower Cambrian sediments of the Sichuan Basin, SW China. *Geol. J.* **2019**, *55*, 3083–3096. [CrossRef]

32. Chen, D.Z.; Wang, J.G.; Qing, H.R.; Yan, D.T.; Li, R.W. Hydrothermal venting activities in the Early Cambrian, South China: Petrological, geochronological and stable isotopic constraints. *Chem. Geol.* **2009**, *258*, 168–181. [CrossRef]

33. Jia, Z.B.; Hou, D.J.; Sun, D.Q.; Jiang, Y.H.; Zhao, Z.; Zhang, Z.M.; Hong, M.; Chang, Z.; Dong, L.C. Characteristics and geological implications of rare earth elements in black shale in hydrothermal sedimentation areas: A case study from the Lower Cambrian Niutitang Fm shale in central and eastern Guizhou. *Nat. Gas Ind.* **2018**, *38*, 44–51. (In Chinese with English abstract)

34. Li, C.; Shi, W.; Cheng, M.; Jin, C.S.; Algeo, T.J. The redox structure of Ediacaran and early Cambrian oceans and its controls. *Sci. Bull.* **2020**, *65*, 2141–2149. [CrossRef]

35. Awan, R.S.; Liu, C.L.; Gong, H.W.; Dun, C.; Tong, C.; Chamssidini, L.G. Paleo-sedimentary environment in relation to enrichment of organic matter of Early Cambrian black rocks of Niutitang Formation from Xiangxi area China. *Mar. Pet. Geol.* **2020**, *112*, 104057. [CrossRef]

36. Wu, Y.W.; Tian, H.; Gong, D.J.; Li, T.F.; Zhou, Q. Paleo-environmental variation and its control on organic matter enrichment of black shales from shallow shelf to slope regions on the Upper Yangtze Platform during Cambrian Stage 3. *Palaeogeogr. Palaeoclimatol. Palaeoecol.* **2020**, *545*, 109653. [CrossRef]

37. Gao, Z.Y.; Liang, Z.; Hu, Q.H.; Jiang, Z.X.; Xuan, Q.X. A new and integrated imaging and compositional method to investigate the contributions of organic matter and inorganic minerals to the pore spaces of lacustrine shale in China. *Mar. Pet. Geol.* **2021**, *127*, 104962. [CrossRef]

38. Ning, S.T.; Xia, P.; Hao, F.; Tian, J.Q.; Zhong, Y.; Zou, N.N.; Fu, Y. Shale facies and its relationship with sedimentary environment and organic matter of Niutitang black shale, Guizhou Province. *Nat. Gas Geosci.* **2021**, *32*, 1297–1307.

39. Dai, C.G.; Zheng, Q.Q.; Chen, J.S.; Wang, M.; Zhang, H. The metallogenic geological background of the Xuefeng-Caledonian tectonic cycle in Guizhou, China. *Earth Sci. Front.* **2013**, *20*, 219–225.

40. Shu, L.S. An analysis of principal features of tectonic evolution in South China Block. *Geol. Bull. China* **2012**, *31*, 1035–1053. (In Chinese with English abstract)

41. Charvet, J. The Neoproterozoic-Early Paleozoic tectonic evolution of the South China Block: An overview. *J. Asian Earth Sci.* **2013**, *74*, 198–209. [CrossRef]

42. Wang, J.; Li, Z.X. History of Neoproterozoic rift basins in South China: Implications for Rodinia break–up. *Precambrian Res.* **2003**, *122*, 141–158. [CrossRef]

43. Wang, X.Q.; Shi, X.Y.; Jiang, G.Q.; Zhang, W.H. New U-Pb age from the basal Niutitang Formation in South China: Implications for diachronous development and condensation of stratigraphic units across the Yangtze platform at the Ediacaran-Cambrian transition. *J. Asian Earth Sci.* **2012**, *48*, 1–8. [CrossRef]

44. Zhu, M.Y.; Zhang, J.M.; Yang, A.H.; Li, G.; Steiner, M.; Erdtmann, B.D. Sinian-Cambrian stratigraphic framework for shallow-to deep-water environments of the Yangtze Platform: An integrated approach. *Prog. Nat. Sci.* **2003**, *13*, 951–960. [CrossRef]

45. Lehmann, B.; Frei, R.; Xu, L.G.; Mao, J.W. Early Cambrian Black Shale-Hosted Mo-Ni and V Mineralization on the Rifted Margin of the Yangtze Platform, China: Reconnaissance Chromium Isotope Data and a Refined Metallogenic Model. *Econ. Geol.* **2016**, *111*, 89–103. [CrossRef]

46. Xu, L.G.; Mao, J.W. Trace element and C-S-Fe geochemistry of Early Cambrian black shales and associated polymetallic Ni-Mo sulfide and vanadium mineralization, South China: Implications for paleoceanic redox variation. *Ore Geol. Rev.* **2021**, *135*, 104210. [CrossRef]

47. Wang, J.G.; Chen, D.Z.; Wang, D.; Yan, D.T.; Zhou, X.Q.; Wang, Q.C. Petrology and geochemistry of chert on the marginal zone of Yangtze Platform, western Hunan, South China, during the Ediacaran-Cambrian transition. *Sedimentology* **2012**, *59*, 809–829. [CrossRef]

48. Xu, L.G.; Lehmann, B.; Mao, J.W.; Qu, W.J.; Du, A.D. Re-Os age of polymetallic Ni-Mo-PGE-Au mineralization in Early Cambrian black shales of South China-A reassessment. *Econ. Geol.* **2011**, *106*, 511–522.

49. Chen, D.Z.; Zhou, X.Q.; Fu, Y.; Wang, J.G.; Yan, D.T. New U–Pb zircon ages of the Ediacaran-Cambrian boundary strata in South China. *Terra Nova* **2015**, *27*, 62–68. [CrossRef]

50. Mclennan, S.M. Relationships between the trace element composition of sedimentary rocks and upper continental crust. *Geochem. Geophys. Geosystems* **2001**, *2*, 2000GC000109. [CrossRef]

51. Gromet, L.P.; Haskin, L.A.; Korotev, R.L. The "North American shale composite": Its compilation, major and trace element characteristics. *Geochim. Et Cosmochim. Acta* **1984**, *48*, 2469–2482. [CrossRef]

52. Usui, A.; Someya, M. Distribution and composition of marine hydrogenetic and hydrothermal manganese deposits in the northwest Pacific. *Geol. Soc. Lond. Spec. Publ.* **1997**, *119*, 177–198. [CrossRef]

53. Jia, Z.B.; Hou, D.J.; Sun, D.Q.; Huang, Y.X. Hydrothermal sedimentary discrimination criteria and its coupling relationship with the source rock. *Nat. Gas Geosci.* **2016**, *27*, 1025–1034, (In Chinese with English abstract).

54. Zhang, K.; Liu, R.; Liu, Z.J.; Li, L.; Wu, X.P.; Zhao, K.A. Influence of palaeoclimate and hydrothermal activity on organic matter accumulation in lacustrine black shales from the Lower Cretaceous Bayingebi Formation of the Yin'e Basin, China. *Palaeogeogr. Palaeoclimatol. Palaeoecol.* **2020**, *560*, 110007.

55. Yuan, W.; Liu, G.; Bulseco, A.; Xu, L.G.; Zhou, X.X. Controls on U enrichment in organic–rich shales from the Triassic Yanchang Formation, Ordos Basin, Northern China. *J. Asian Earth Sci.* **2021**, *212*, 104735. [CrossRef]

56. Zhu, G.Y.; Wang, P.J.; Li, T.T.; Zhao, K.; Zheng, W.; Feng, X.B.; Shen, J.; Grasby, S.E.; Sun, G.Y.; Tang, S.L.; et al. Mercury record of intense hydrothermal activity during the early Cambrian, South China. *Palaeogeogr. Palaeoclimatol. Palaeoecol.* **2021**, *568*, 110294. [CrossRef]

57. Wei, S.C.; Chen, Q.F.; Fu, Y.; Cui, T.; Liang, H.P.; Ge, Z.H.; Zhang, P.; Zhang, Y. Origin of Cherts during the Ediacaran-Cambrian Transition in Hunan and Guizhou Provinces, China: Evidences from REE and Ge/Si. *Acta Sci. Nat. Univ. Pekin.* **2018**, *54*, 1010–1020. (In Chinese with English abstract)

58. Ruhlin, D.E.; Owen, R.M. The rare earth element geochemistry of hydrothermal sediments from the East Pacific Rise: Examination of a seawater scavenging mechanism. *Geochim. Et Cosmochim. Acta* **1986**, *50*, 393–400. [CrossRef]

59. Nesbitt, H.W.; Young, G.M. Early Proterozoic climates and plate motions inferred from major element chemistry of lutites. *Nature* **1982**, *299*, 715–717. [CrossRef]

60. Nesbitt, H.W.; Young, G.M. Formation and Diagenesis of Weathering Profiles. *J. Geol.* **1989**, *97*, 129–147. [CrossRef]

61. Mclennan, S.M. Weathering and Global Denudation. *J. Geol.* **1993**, *101*, 295–303. [CrossRef]

62. Fedo, C.M.; Wayne Nesbitt, H.; Young, G.M. Unraveling the effects of potassium metasomatism in sedimentary rocks and paleosols, with implications for paleoweathering conditions and provenance. *Geology* **1995**, *23*, 921–924. [CrossRef]

63. Yao, W.H.; Li, Z.X.; Li, W.X. Was there a Cambrian ocean in South China?—Insight from detrital provenance analyses. *Geol. Mag.* **2015**, *152*, 184–191. [CrossRef]

64. Tribovillard, N.; Algeo, T.J.; Lyons, T.; Riboulleau, A. Trace metals as paleoredox and paleoproductivity proxies: An update. *Chem. Geol.* **2006**, *232*, 12–32. [CrossRef]

65. Rachold, V.; Brumsack, H.J. Inorganic geochemistry of Albian sediments from the Lower Saxony Basin NW Germany: Palaeoenvironmental constraints and orbital cycles. *Palaeogeogr. Palaeoclimatol. Palaeoecol.* **2001**, *174*, 121–143.

66. Wang, Z.G.; Yu, X.Y.; Zhao, Z.H. *Geochemistry of Rare Earth Elements*; Science Press: Beijing, China, 1989.

67. Wright, J.; Schrader, H.; Holser, W.T. dox variations in ancient oceans recorded by rare earth elements in fossil apatite. *Geochim. Cosmochim. Acta* **1987**, *51*, 631–644. [CrossRef]

68. Kechiched, R.; Laouar, R.; Bruguier, O.; Kocsis, L.; Salmi-Laouar, S.; Bosch, D.; Ameur-Zaimeche, O.; Foufou, A.; Larit, H. Comprehensive REE + Y and sensitive redox trace elements of Algerian phosphorites (Tébessa, eastern Algeria): A geochemical study and depositional environments tracking. *J. Geochem. Explor.* **2020**, *208*, 106396. [CrossRef]

69. Zhao, C.J.; Kang, Z.H.; Hou, Y.H.; Yu, X.D.; Wang, E.B. Geochemical characteristics of rare earth elements and their geological significance of Permian shales in Lower Yangtze Area. *Earth Sci.* **2020**, *45*, 4118–4127. (In Chinese with English abstract)

70. Schoepfer, S.D.; Shen, J.; Wei, H.Y.; Tyson, R.V.; Ingall, E.; Algeo, T.J. Total organic carbon, organic phosphorus, and biogenic barium fluxes as proxies for paleomarine productivity. *Earth Sci. Rev.* **2015**, *149*, 23–52. [CrossRef]

71. Moore, J.K.; Fu, W.W.; Primean, F.; Britten, G.L.; Lindsay, K.; Long, M.; Doney, S.C.; Mahowald, N.; Hoffman, F.; Randerson, J.T. Sustained climate warming drives declining marine biological productivity. *Science* **2018**, *359*, 1139–1143. [CrossRef]

72. Chen, G.; Gang, W.Z.; Chang, X.C.; Wang, N.; Zhang, P.F.; Cao, Q.Y. Paleoproductivity of the Chang 7 unit in the Ordos Basin (North China) and its controlling factors. *Palaeogeogr. Palaeoclimatol. Palaeoecol.* **2020**, *551*, 109741. [CrossRef]

73. House, B.M.; Norris, R.D. Unlocking the barite paleoproductivity proxy: A new high–throughput method for quantifying barite in marine sediments. *Chem. Geol.* **2020**, *552*, 119664. [CrossRef]

74. Lu, S.F.; Zhang, M.; Zhong, N.N. *Oil and Gas Geochemistry*; Petroleum Industry Press: Beijing, China, 2008.

75. Dymond, J.; Suess, E.; Lyle, M. Barium in Deep–Sea Sediment: A Geochemical Proxy for Paleoproductivity. *Paleoceanography* **1992**, *7*, 163–181. [CrossRef]

76. Wang, R.Y.; Hu, Z.Q.; Nie, H.K.; Liu, Z.B.; Chen, Q.; Gao, B.; Liu, G.X.; Gong, D.J. Comparative analysis and discussion of shale reservoir characteristics in the Wufeng-Longmaxi and Niutitang formations. *Pet. Geol. Exp.* **2018**, *40*, 639–649. (In Chinese with English abstract)

77. Li, D.L.; Li, R.X.; Tan, C.Q.; Zhao, D.; Xue, T.; Zhao, B.S.; Khaled, A.; Liu, F.T.; Xu, F. Origin of silica, paleoenvironment, and organic matter enrichment in the Lower Paleozoic Niutitang and Longmaxi formations of the northwestern Upper Yangtze Plate: Significance for hydrocarbon exploration. *Mar. Pet. Geol.* **2019**, *103*, 404–421. [CrossRef]
78. Ross, D.J.K.; Bustin, R.M. Investigating the use of sedimentary geochemical proxies for paleoenvironment interpretation of thermally mature organic–rich strata: Examples from the Devonian–Mississippian shales, Western Canadian Sedimentary Basin. *Chem. Geol.* **2009**, *260*, 1–19. [CrossRef]
79. Schenau, S.J.; Reichart, G.J.; Lange, G. Phosphorus burial as a function of paleoproductivity and redox conditions in Arabian Sea sediments. *Geochim. Cosmochim. Acta* **2005**, *69*, 919–931. [CrossRef]
80. Shen, J.; Schoepfer, S.D.; Feng, Q.L.; Zhou, L.; Yu, J.X.; Song, H.Y.; Wei, H.Y.; Algeo, T.J. Marine productivity changes during the end–Permian crisis and Early Triassic recovery. *Earth Sci. Rev.* **2015**, *149*, 136–162. [CrossRef]
81. Jones, B.; Manning, D.A.C. Comparison of geochemical indices used for the interpretation of palaeoredox conditions in ancient mudstones. *Chem. Geol.* **1994**, *111*, 111–129. [CrossRef]
82. Algeo, T.J.; Tribovillard, N. Environmental analysis of paleoceanographic systems based on molybdenum–uranium covariation. *Chem. Geol.* **2009**, *268*, 211–225. [CrossRef]
83. Kurek, M.R.; Harir, M.; Shukle, J.T.; Schroth, A.W.; Schmitt-Kopplin, P.; Druschel, G.K. Seasonal transformations of dissolved organic matter and organic phosphorus in a polymictic basin: Implications for redox-driven eutrophication. *Chem. Geol.* **2021**, *573*, 120212. [CrossRef]
84. Severmann, S.; Anbar, A. Reconstructing Paleoredox Conditions through a Multitracer Approach: The Key to the Past Is the Present. *Elements* **2009**, *5*, 359–364. [CrossRef]
85. Algeo, T.J.; Li, C. Redox classification and calibration of redox thresholds in sedimentary systems. *Geochim. Cosmochim. Acta* **2020**, *287*, 8–26. [CrossRef]
86. Bau, M.; Dulski, P. Distribution of yttrium and rare-earth elements in the Penge and Kuruman iron–formations, Transvaal Supergroup, South Africa. *Precambrian Res.* **1996**, *79*, 37–55.
87. Wignall, P.B.; Twitchett, R.J. Oceanic anoxic and the end Permian mass extinction. *Science* **1996**, *272*, 1155–1158. [CrossRef]
88. Zhang, M.L.; Guo, W.; Shen, J.; Liu, K.; Zhou, L.; Feng, Q.L.; Lei, Y. New progreses on geochemical indicators of ancient oceanic redox condition. *Geol. Sci. Technol. Inf.* **2017**, *36*, 95–106, (In Chinese with English abstract).
89. Algeo, T.J.; Liu, J.S. A re-assessment of elemental proxies for paleoredox analysis. *Chem. Geol.* **2020**, *540*, 119549. [CrossRef]
90. Elderfield, H.; Greaves, M.J. The rare earth elements in seawater. *Nature* **1982**, *296*, 214–219. [CrossRef]
91. Wignall, P.B.; Newton, R.; Brookfield, M.E. Pyrite framboid evidence for oxygenpoor deposition during the Permian-Triassic crisis in Kashmir. *Palaeogeogr. Palaeoclimatol. Palaeoecol.* **2005**, *216*, 183–188. [CrossRef]
92. Zhou, C.M.; Jiang, S.Y. Palaeoceanographic redox environments for the lower Cambrian Hetang Formation in South China: Evidence from pyrite framboids, redox sensitive trace elements, and sponge biota occurrence. *Palaeogeogr. Palaeoclimatol. Palaeoecol.* **2009**, *271*, 279–286. [CrossRef]
93. Helz, G.R.; Miller, C.V.; Charnock, J.M.; Mosselmans, J.F.W.; Pattrick, R.A.D.; Garner, C.D.; Vaughan, D.J. Mechanism of molybdenum removal from the sea and its concentration in black shales: EXAFS evidence. *Geochim. Cosmochim. Acta* **1996**, *60*, 3631–3642. [CrossRef]
94. Rowe, H.D.; Loucks, R.G.; Ruppel, S.C.; Rimmer, S.M. Mississippian Barnett Formation, Fort Worth Basin, Texas: Bulk geochemical inferences and Mo–TOC constraints on the severity of hydrographic restriction. *Chem. Geol.* **2008**, *257*, 16–25. [CrossRef]
95. Tribovillard, N.; Algeo, T.J.; Baudin, F.; Riboulleau, A. Analysis of marine environmental conditions based onmolybdenum-uranium covariation-Applications to Mesozoic paleoceanography. *Chem. Geol.* **2012**, *324–325*, 46–58. [CrossRef]
96. Algeo, T.J.; Rowe, H. Paleoceanographic applications of trace-metal concentration data. *Chem. Geol.* **2012**, *324–325*, 6–18. [CrossRef]

Article

Physical Similarity Simulation of Deformation and Failure Characteristics of Coal-Rock Rise under the Influence of Repeated Mining in Close Distance Coal Seams

Pengze Liu [1], Lin Gao [1,2,3,4,*], Pandong Zhang [1], Guiyi Wu [1,3], Yongyin Wang [1,3], Ping Liu [1,3], Xiangtao Kang [1,3], Zhenqian Ma [1,3], Dezhong Kong [1,3] and Sen Han [1,3]

[1]   College of Mining, Guizhou University, Guiyang 550025, China; liupengze0205@163.com (P.L.); zpdznhy@163.com (P.Z.); gywu@gzu.edu.cn (G.W.); yywang2@gzu.edu.cn (Y.W.); pliu1@gzu.edu.cn (P.L.); xiaokangedu@163.com (X.K.); zqma@gzu.edu.cn (Z.M.); dzkong@gzu.edu.cn (D.K.); shan1@gzu.edu.cn (S.H.)
[2]   Coal Mine Roadway Support and Disaster Prevention Engineering Research Center, Beijing 100083, China
[3]   National & Local Joint Laboratory of Engineering for Effective Utilization of Regional Mineral Resources from Karst Areas, Guiyang 550025, China
[4]   Key Laboratory of Mining Disaster Prevention and Control, Qingdao 266590, China
*   Correspondence: lgao@gzu.edu.cn

Citation: Liu, P.; Gao, L.; Zhang, P.; Wu, G.; Wang, Y.; Liu, P.; Kang, X.; Ma, Z.; Kong, D.; Han, S. Physical Similarity Simulation of Deformation and Failure Characteristics of Coal-Rock Rise under the Influence of Repeated Mining in Close Distance Coal Seams. *Energies* **2022**, *15*, 3503. https://doi.org/10.3390/en15103503

Academic Editors: Jing Li, Yidong Cai, Lei Zhao and Manoj Khandelwal

Received: 9 March 2022
Accepted: 5 May 2022
Published: 11 May 2022

Publisher's Note: MDPI stays neutral with regard to jurisdictional claims in published maps and institutional affiliations.

Abstract: Aiming at the problem that it is difficult to achieve accurate laying of model and precise excavation of roadways in special surrounding rock structure roadway according to conventional physical similarity simulation, which reduces the reliability of experimental results. An accurate laying of model and precise excavation of roadway method, named "labeling positioning and drawing line, presetting roadway model" (LPDLPRM), was proposed. The physical similarity simulation of deformation and failure characteristics of surrounding rock of coal-rock rise, under the influence of repeated mining in close distance coal seams, was carried out based on the method and infrared detection. The results show that the coal-rock rise in close distance coal seams was affected by repeated mining disturbances, and the surrounding rock of coal-rock rise was characterized by obvious asymmetric deformation, specific for the stress and strain near the coal pillar were higher than that of other parts, and cracks near the coal pillar were denser than other parts; when the coal seam is mined in which the coal-rock rise is located, the stress concentration of the surrounding rock near the rise was weakened by mining pressure relief in the upper coal seam; the stress concentration of the surrounding rock near the rise increases when the coal and the lower coal seam are mined, and the stress on the right side (coal pillar side) near the coal-rock rise was the most concentrated. Therefore, it is important to take measures to strengthen support near the coal pillar and to control asymmetric deformation when the coal-rock rise is influenced by repeated mining.

Keywords: similar simulation test; precise excavation of roadway; repeated mining; coal-rock roadway; infrared thermal image

1. Introduction

Minable coal seams in Guizhou Province of China are mainly characterized by complex geological structure, close distance, and thin and medium thick coal seams [1,2]. Compared with a single thick coal seam, the mining process in Guizhou is mostly accompanied by violent mine pressure and continuous large deformation of surrounding rock [3]. These are the important factors to restrict the improvement of the coal mine safety situation and production efficiency in Guizhou. Therefore, it is urgent to further reveal the deformation characteristics of surrounding rock and to provide the scientific basis for support design in Guizhou.

In general, the commonly used research methods in mining engineering include theoretical analysis, numerical simulation, physical similarity simulation, etc. [4,5], while physical similarity simulation is widely used in the field of mining engineering, because it has some obvious advantages, such as intuitive test effect, short test cycle, high simulation degree and easy operation [6–13].

For example, Lin et al. [14] simulated roadway excavation in a coal mine, used a self-made model to build a test model, which was controlled by microcomputer and employing an electro-hydraulic servo universal testing machine to load. Chen et al. [15] simulated roadway excavation with high geostress, which was supported by an anchor used self-made true three-dimensional loading model test system. Pornkasem Jongpradist et al. [16] studied the fracture mechanism of surrounding rock for high internal pressure roadway through establishing physical similarity model. Yan et al. [17] studied the dynamic failure process of surrounding rock from stability to instability for a horizontal extra thick coal seam through physical similarity simulation test, and revealed the stress and displacement characteristics of a fully mechanized mining roadway. Hui et al. [18] adopted a manual opening model center to simulate tunnel excavation in a physical similarity simulation test. Mishra Swapnil et al. [19] studied the behavior and pattern of tunnel damage subjected to the different dynamic loading conditions by a similar simulation. Shan et al. [20] adopted the self-developed new simulation test device of roadway support, which can apply dynamic and static load, and studied the influence of dynamic and static load on roadway deformation law by comparison. Chang et al. [21] reproduced a high stress environment of deep roadway through applying different pressure on roof, floor and both sides of the roadway with a four-way loading simulation test bench. Wook and Lee [22] investigated pile load distribution and ground behavior due to tunnelling below a grouped pile using a laboratory model test. Li et al. [23] studied the failure characteristics of a roadway under uniaxial and biaxial compression using a self-developed biaxial loaded roadway simulation test system equipped with a small and medium-sized motor. Yuan et al. [24] proposed a new method of similar material simulation test combined with orthogonal test and multiple regression analysis, and studied the protective effect of upper protective layer mining of a steeply inclined coal seam with variable interval using this method. Wang et al. [25] explored the dynamic response and impact failure characteristics of coal and rock mass under a steeply inclined extra-thick coal seam through the physical simulation experiment with static and dynamic coupling loading. Idinger Gregor et al. [26] investigated some aspects of the collapse at tunnel face for different overburden pressures on a small-scale tunnel model in a geotechnical centrifuge. He et al. [27] committed to the simulation of a roadway excavation in the geologically horizontal strata at great depth based on a physical modeling test. Li et al. [28] investigated the evolution laws of floor rock fracture under the mining dynamic loading effects by a similar simulation test. Ren et al. [29] studied the characteristics of the breaking process of overburden rocks in shallow coal workface through physical similarity simulation. Shi et al. [30] studied the movement law, failure mechanism and fracture evolution of overlying strata of gob-side entry driving in thick coal seam by combining physical model test and numerical simulation. Berthoz Nicolas et al. [31] studied mechanisms of face collapse and face blow-out of tunnels driven in soft grounds with pressurized shield tunnel boring machine with original laboratory reduced-scale model. Feng et al. [32] studied the failure law of rock during coal mining by physical similar material simulation experiment. Xue et al. [33] simulated the movement and fracture evolution of the overlying strata after the coal seam is extracted. Yang et al. [34] studied the development of mining stress and the effect of large height upward mining pressure relief gas using physical similarity simulation. Zhang et al. [35] analyzed the displacement, strain and vertical stress field of surrounding rock near the fault, and determined the influence of coal pillar width by similar simulation combined with digital speckle.

Summarizing above literature, it was found that most of the existing studies on the deformation characteristics of roadway surrounding rock and pressure behavior using physical similarity simulation focused on roadways in nearly horizontal thick coal seam. However, coal-rock rise widely used in close distance, thin and medium thick coal seams in Guizhou Province of China. Due to the heterogeneity of surrounding rock structure and the existence of the coal–rock interface, the mechanical properties of the surrounding rock were influenced, and it was difficult to achieve accurate laying of model and precise excavation of roadways. Though physical similarity simulation is widely used to study the mine pressure behavior of roadways, the relevant literature is rarely mentioned on close distance, thin and medium thick coal seams. To further intuitive reveal the deformation characteristics of coal-rock rise under the influence of repeated mining in close distance coal seams, a physical similarity simulation test method of accurate laying of model and precise excavation of roadway was proposed. Combined with infrared thermal detection means, physical similar simulation experiment research was conducted for the roadways of special surrounding rock structures.

2. The Proposition of Accurate Laying and Excavation Method of Similarity Simulation Model in Roadway of Special Surrounding Rock Structure

Special surrounding rock structures such as inclined coal and rock strata and coal-rock roadways were common in Guizhou Province of China [36–38]. It was difficult to achieve precise laying of the conventional physical similarity test model and precise excavation of roadway in these coal and rock strata. That resulted in the physically similar experiment results that were not accurate, and its application was limited. Based on this, the research group proposed an accurate laying of model and precise excavation of roadway test method named "labeling positioning and drawing line, presetting roadway model" (LPDLPRM).

2.1. Basic Process of the Test Method

This test method mainly consists of determining similar materials and its ratio, model design, installing and debugging test bench, mixing similar materials, making roadway model, labeling positioning and drawing line, laying model, presetting roadway model, maintaining model and others. Among these, making roadway model, labeling positioning and drawing line, and presetting roadway model are the core contents of this method. The detailed process are as follows:

1. Making roadway model.

The roadway model is the core component of the accurate laying of model and precise excavation of roadway method. The roadway model consists of an inner frame produced by Φ 5 mm steel and PC plastic sheet surrounding the inner frame. In order to ensure the smooth pull out of the model from coal and rock strata and to reduce the large friction when roadway excavation is conducted in the later stage of the test, the PC plastic sheet is brushed with oil. This model is shown in Figure 1.

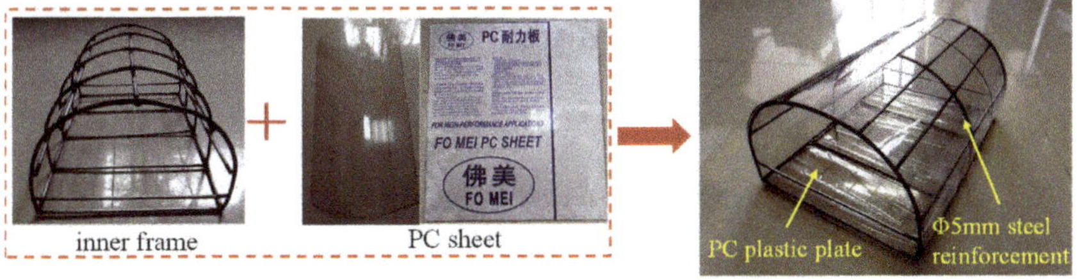

Figure 1. Composition of roadway model.

2. Labeling positioning and drawing line.

According to the model design drawing, the laying position of each coal and rock strata on the baffle plate of the test bench are drawn. Additionally, the positioning prompt labels of roadway model and test components such as pressure box are affixed on corresponding position where both sides and the baffle at the back of the model. That can ensure the accurate laying of subsequent all coal and rock strata and roadway model. Then, labeling the names of all coal and rock strata, and drawing the boundary line of each coal and rock strata, drawing the position of each stress sensors and the outline of the roadway.

3. Presetting roadway model.

When the model is laid to the corresponding labels of the outline of the roadway, the roadway model is embedded according to the calibration position, the pressure boxes are buried according to the positioning labels, and a layer of edible oil is brushed on the outside of the roadway model. After the laying and maintenance are completed, the roadway model is pulled out to simulate the excavation of the roadway.

2.2. The Advantages of this Experimental Method

Compared to the traditional experimental method, through the labeling the roadway outline and coal and rock boundary in advance, this method can realize not only the accurate laying of model in coal and rock strata (especially the inclined coal and rock strata), but also the precise excavation of roadway. At the same time, through the position labels, pressure sensors and others can be accurately embedded to ensure the location of pressure sensors consistent with the design scheme, so that the test data collected is highly reliable and accurate. This method aims to solve the problems of the uneven thickness of coal and rock strata, the irregular outline of excavated roadway and unreliability monitoring data of the physical similarity simulation experiment in special surrounding rock structure.

3. Overview of the Project

A coal mine in Guizhou Province of China is characterized by complicated geological structures where faults and folds widely existed. Its minable coal seams are typical close distance coal seams, which roof and floor are mostly mudstone, argillaceous siltstone and other soft rocks. Rise for transportation is located in 17# coal and its floor, and about two-thirds of which cross section is in 17# coal. The roadway is driven along the roof of 17# coal. The average thickness and dip angle of 17# coal are 2.5 m and 20°, respectively. The cross section of the roadway is a straight wall and semicircular arch with a lower width of 5.5 m and middle height of 3.3 m. It is a typical coal-rock roadway. The position relation of each coal and rock strata and rise for transportation is shown in Figure 2.

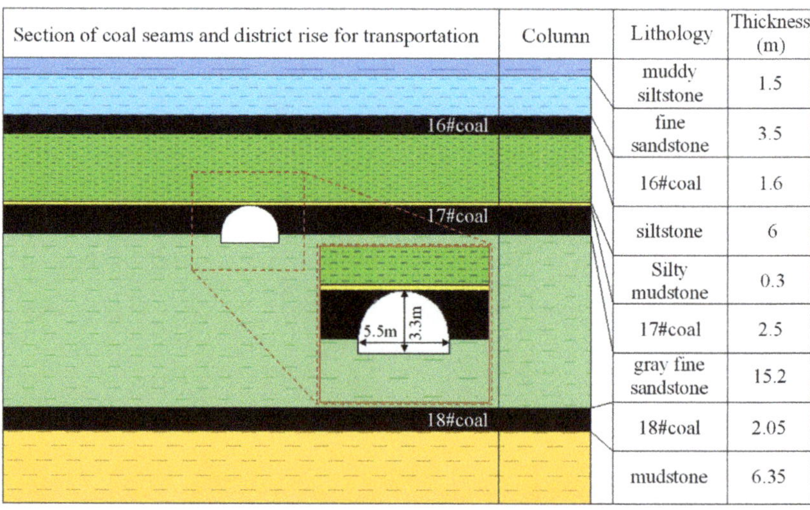

Section of coal seams and district rise for transportation	Column	Lithology	Thickness (m)
		muddy siltstone	1.5
		fine sandstone	3.5
		16#coal	1.6
		siltstone	6
		Silty mudstone	0.3
		17#coal	2.5
		gray fine sandstone	15.2
		18#coal	2.05
		mudstone	6.35

Figure 2. Strata histogram in this research.

4. Experiment Plan of the Physical Similarity Simulation

4.1. Test Equipment and Test System

The test device is the QKX-EW-2 two-dimensional physical similarity simulation test bench (Figure 3) with a length of 3000 mm, width of 300 mm and height of 1800 mm. There are 10 vertical loading points, and the maximum pressure/load that can be applied is 1 MPa/1000 kN. Three lateral loading points are arranged on the left sides and three on the right sides with maximum loading pressure/load being 1 MPa/600 kN for each loading point, and pressure of the loading system is 15 MPa. The loading plate is pushed by the oil cylinder to realize vertical and lateral loading. Among them, the dimensions of the 10 vertical active loading plates are all 300 mm × 300 mm (length × width). The dimensions of the three lateral loading plates on the left or right sides are 500 mm × 300 mm (height × width), and the rated voltage/power is 380 V/55 kW. They can be used at −10–65 °C. At the same time, in order to reduce the influence of boundary effect, the excavation area is arranged near the diagonal of the test bench as far as possible.

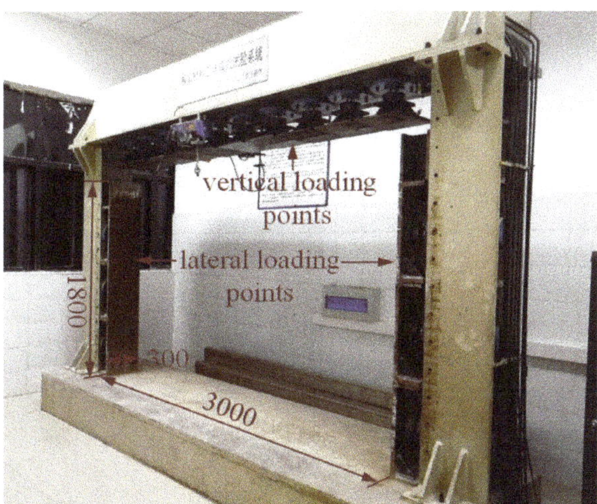

Figure 3. QKX-EW-2 two-dimensional physical similarity simulation test bench.

The high-definition and high-speed digital shooting system was adopted for data collection. Using CanonEOS750D high-definition digital camera controlled by mobile phone through wireless connection to track, shoot and record changes of surrounding rock cracks at key moments. Using high-speed camera Qianyanlang 2F04 equipped with EF-200LED lighting to capture the instantaneous cracks and deformation of the surrounding rock. Using SONY full HD digital camera to record the whole process of the experiment and to ensure every detail of experiment to be recorded.

As the relationship between surrounding rock stress, strain and surrounding rock surface infrared radiation can be expressed qualitatively. Additionally, with the increase of load, the infrared radiation temperature is higher [39–45], Fluke Ti450PRO infrared thermal imager was used for infrared detection of roadway surrounding rock during the experiment. The temperature measurement range of Fluke Ti450PRO infrared thermal imager is −20–1200 °C, the thermal sensitivity is 0.025 °C (30 Mk), and the image resolution is 640 × 480. The principle of Fluke Ti450PRO infrared thermal imager is to passively receive the infrared radiation (heat) from the measured target through non-contact nondestructive detection and to convert this heat into a visual image (IR image) with temperature data, which is shown in Figure 4. The image data were imported into SmartView4.3 professional infrared thermal analysis software (Figure 5) for data processing and analysis, then the two-dimensional and three-dimensional temperature fields of the detection object were obtained, which can be used to analyze the stress state of the surrounding rock indirectly.

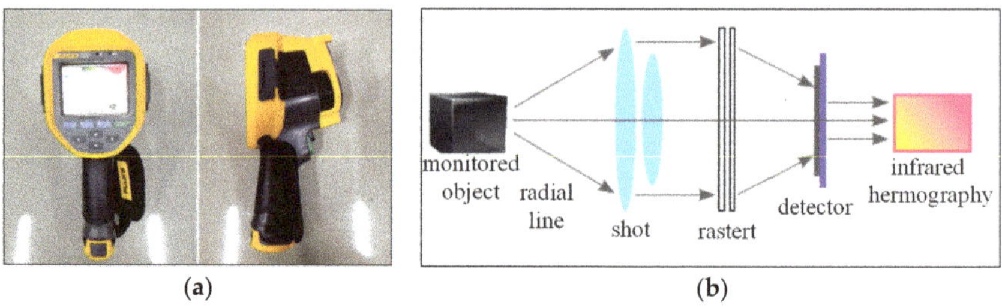

(**a**)   (**b**)

Figure 4. Infrared thermal imager and its working principle. (**a**) Infrared thermal imager; (**b**) and working principle.

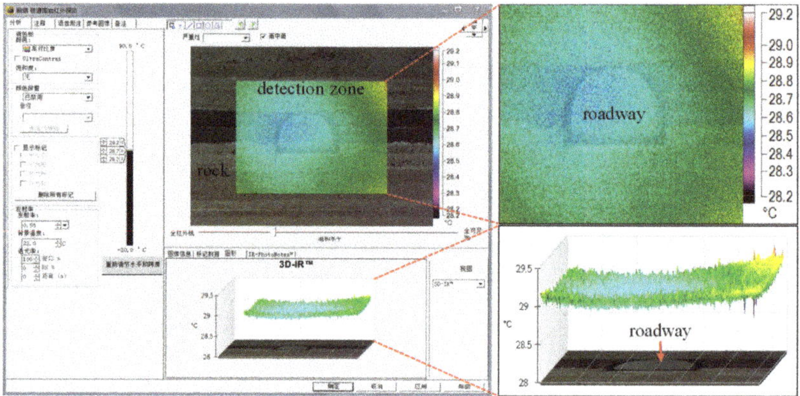

Figure 5. Infrared thermal analysis software. (Its function keys only provide Chinese display).

4.2. Experimental Materials and Making Similar Models

1. Determination of similar materials and ratio number. The selection of similar materials and the determination of reasonable ratio number are crucial to obtain accurate and reliable results of similar simulation experiment. According to the mechanical parameters of actual coal and rock strata and similarity theory, the geometric similarity ratio of the model $C_l = 25$, the density similarity ratio $C_\gamma = 1.5$, the stress similarity ratio $C_p = 37.5$, and the time similarity ratio $C_t = 5$ were chosen. Similar materials (Figure 6) choose sand as aggregate, lime and gypsum as cementing materials, and mica powder was spread between layers of coal and rock to simulate bedding. According to the calculation method of the strength value of coal and rock mass simulated by similar materials, the strength test of similar materials with different proportions was carried out for the main coal and rock strata. After repeated adjustment, the reasonable ratio number of similar materials for each coal and rock strata was obtained, as shown in Table 1.

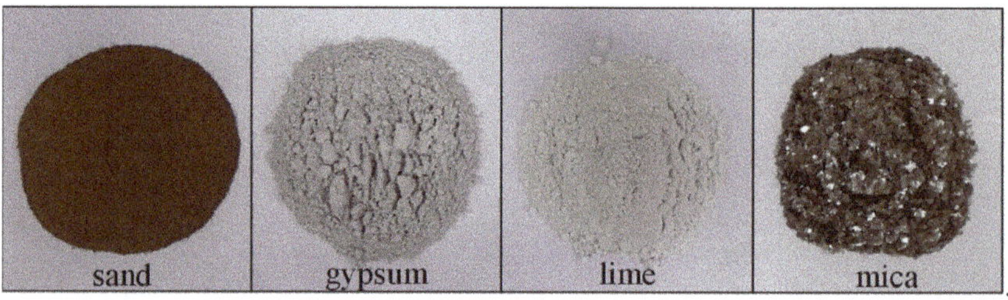

Figure 6. Main similar materials.

Table 1. Similar material ratio and mechanical parameters.

Rock Stratum Number	Thickness (m)	Lithology	Proportion Number	Model Thickness (cm)	Density (g/cm^3)	Lamination Thickness (cm)	Layer Number	Weight of Water	Compressive Strength (MPa)
9	1.5	muddy siltstone	355	6	1.58	2.0	3	1/9	0.63
8	3.5	fine sandstone	473	14	1.55	2.0	7	1/9	0.57
7	1.6	16#coal	773	6.4	0.93	2.0	3	1/9	0.08
6	6	siltstone	373	24	1.70	2.0	12	1/9	0.75
5	0.3	silty mudstone	582	1.2	1.46	1.2	1	1/9	0.51
4	2.5	17#coal	773	10	0.93	2.0	5	1/9	0.08
3	15.2	gray fine sandstone	472	60.8	1.55	2.0	30	1/9	0.56
2	2.05	18#coal	773	8.2	0.93	2.0	4	1/9	0.08
1	6.35	mudstone	573	25.4	1.42	2.0	13	1/9	0.46

2 Model design. According to the size of roadway, the true thicknesses of each coal and rock strata, the geometric similarity ratio, the size of test bench, and others, the size of the roadway model and its thicknesses of each coal and rock strata were determined, then the model design diagram was drawn. Before the model was laid, the quality of sand and lime in each layer was calculated according to the real thicknesses of each coal and rock strata, geometric similarity ratio, model test bench size, similar material density and ratio number. Then, the preparation of weighing ratio and mixing of similar materials was completed.

3 LPDLPRM. The positioning labels of the roadway model and pressure box were put on the test bench according to the drawing, the separation lines of the coal and rock strata were drawn, mark the names of the coal seam and rock strata. Then, the coal seams and rock strata were laid, respectively, using similar materials which had been matched. The coal seams and rock strata were laid one by one and compacted every 5 cm. In order to make the model layered obviously, a layer of mica powder was evenly spread on it as a weak layer after each strata of coal and rock was laid. When the model was laid to the marked position, the roadway model that has been made was embedded to the marked position. We continued these procedures until the model laying was completed. The test process is shown in Figure 7.

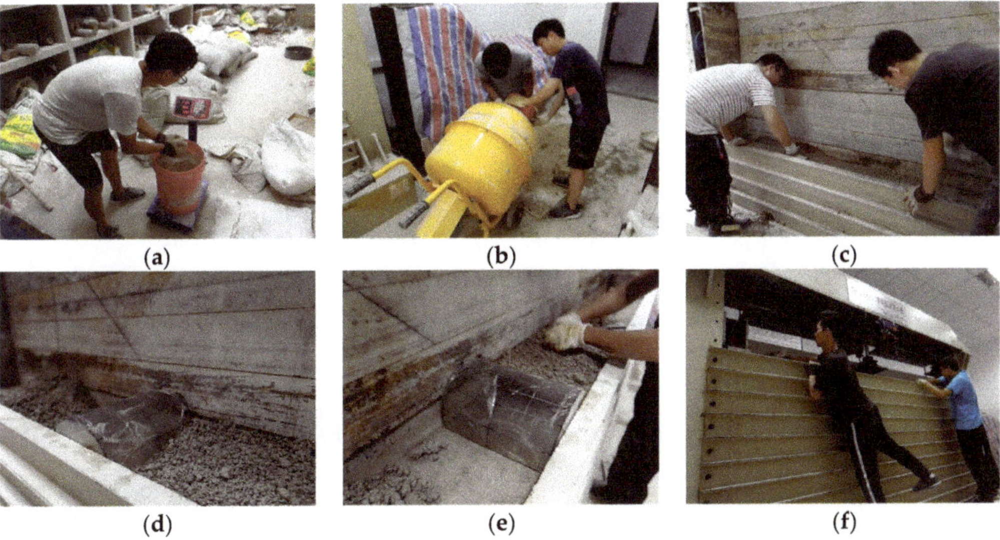

Figure 7. The process of making model. (**a**) Weighing ratio; (**b**) mixing materials; (**c**) laying model; (**d**) presetting of roadway model; (**e**) compacted materials; and (**f**) laying upper part of model.

4   Maintenance of model. Three days after the model was laid, the fixed steel plates in front and back of the model were removed. When the whole model was dry, there were no water marks, and the whole was gray, pressing hard on its surface will not produce obvious depression, indicating that the experimental requirements were met. Then, the dense observation grids were drawn to ensure accurately record the variation characteristics of surrounding rock cracks during roadway excavation. Thus far, the test model was completed, which is shown in Figure 8.

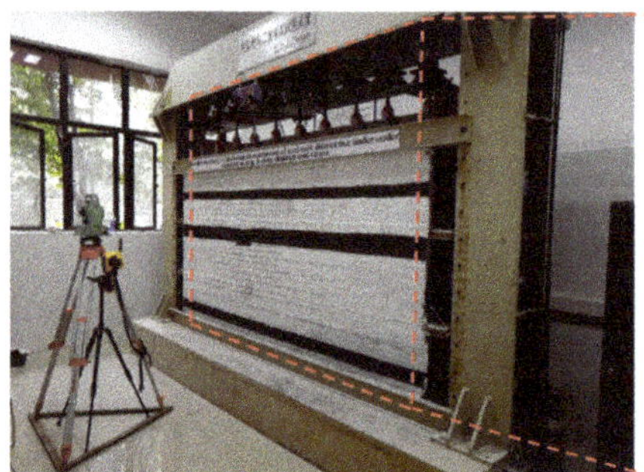
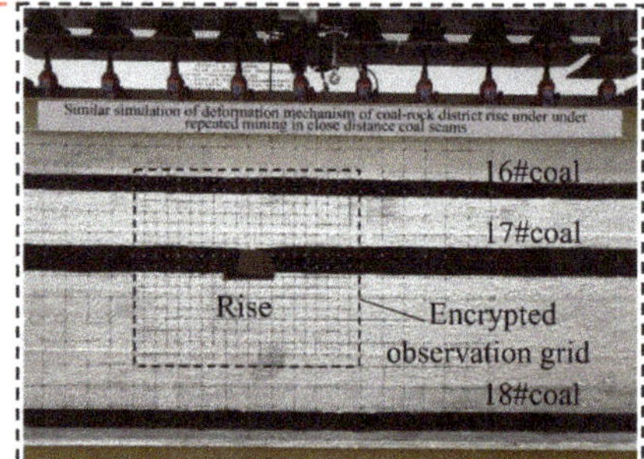

Figure 8. Final drawing of test model.

4.3. Process of the Experiment

The experiment was started when the model was static air-dried to achieve the desired effect and the model strength reached the design requirements. According to the buried depth and stress similarity ratio of the roadway, evenly and gradually vertical load applied on the top of the model, kept the oil pressure of each loading plate stabilized, and kept the load constant, then carried out the roadway excavation (the preset roadway model was extracted). The excavation position was in the middle of coal–rock combination body of 17# coal and lower rock. During the process of the roadway excavation, a high-definition camera was used to capture the instantaneous cracks and deformation of the surrounding rock, and a camera was used to record the whole test until the surrounding rock deformation tended to be stable.

5. Experiment Results and Analysis

5.1. Evolution Characteristics of Surrounding Rock Cracks

During the experiment, roadway excavation was simulated by extracting the preset roadway model, and the minable 16#, 17# and 18# coal seams were simulated by this kind of manual excavation. At the same time, the high-definition digital camera wireless connected to mobile phone was used to track and record the fracture situation of surrounding rock at key moment. The obtained images were imported into AutoCAD in raster image format to obtain the sketch map of fracture distribution of surrounding rock. The results show that after the 16# coal was mined, only a few fine cracks were generated away from the roadway; when the 17# coal (the coal seam in which the coal-rock rise was) was mined, a large range of cracks were produced on the roof and the upper right of the roadway; when the 18# coal was mined, the cracks on the right side of the roadway floor continued to extend downward. The distribution of cracks in surrounding rock of the roadway was asymmetric in the form of left thin and right dense, and the surrounding rock of roadway showed obvious asymmetric deformation characteristics. The crack distribution and evolution sketch of surrounding rock of the coal-rock roadway during the experiment were shown in Figure 9.

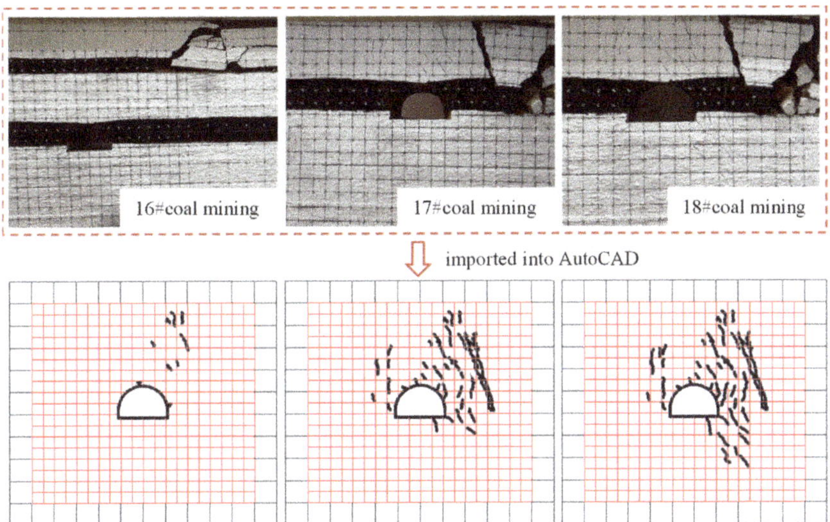

Figure 9. Sketch of fracture distribution and evolution of surrounding rock of coal-rock dstrict rise.

5.2. Analysis of the Results of Infrared Thermal Imager Detection

Under the influence of mined and non-mined roadway surrounding rock for different coal seams (16#, 17#, 18#), the plane temperature field of roadway surrounding rock, large-scale three-dimensional temperature field of roadway surrounding rock, and the temperature changes at different positions of roadway surrounding rock in different coal seam mining were shown in Figures 10–12.

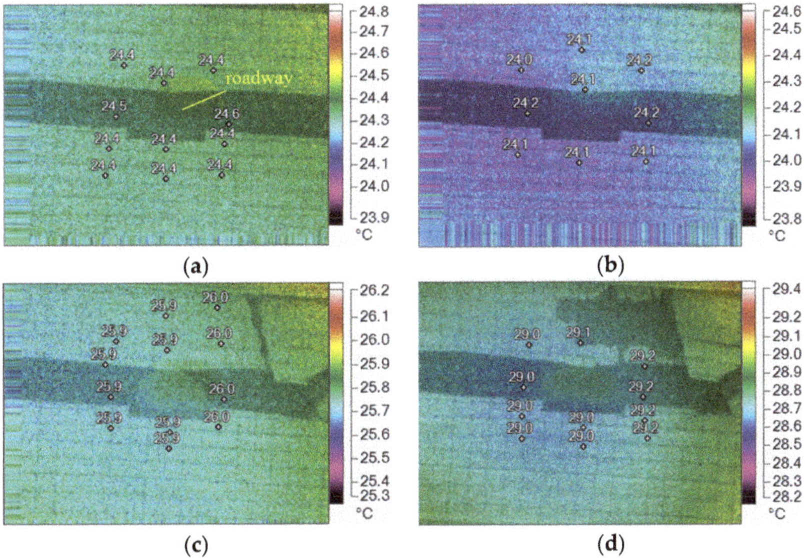

Figure 10. Plane temperature field of surrounding rock in mining roadway in different coal seams. (**a**) No-mining; (**b**) 16# coal mining; (**c**) 17# coal mining; and (**d**) 18# coal mining.

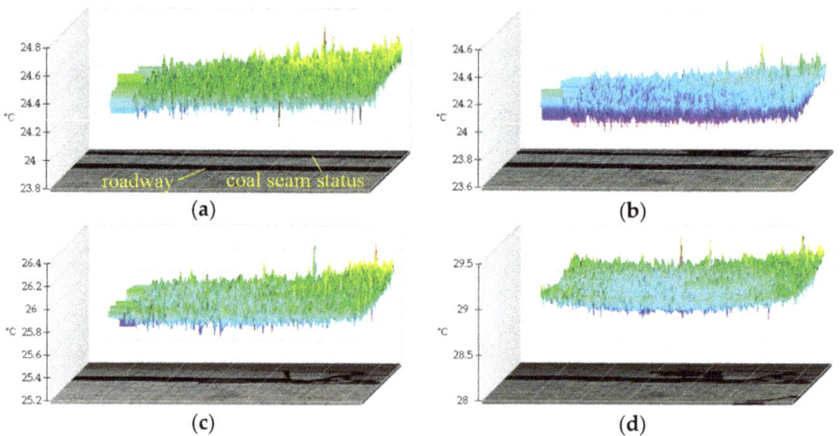

Figure 11. Three-dimensional temperature field of surrounding rock in mining roadway in different coal seams. (**a**) No-mining; (**b**) 16# coal mining; (**c**) 17# coal mining; and (**d**) 18# coal mining.

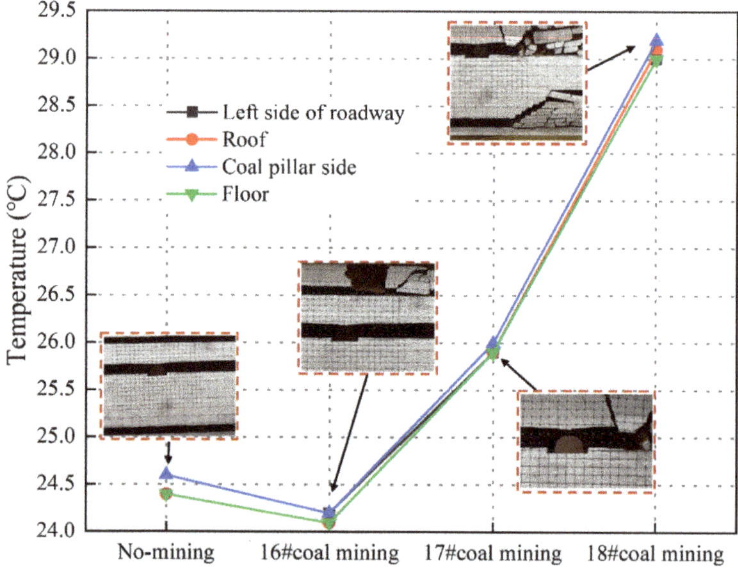

Figure 12. The temperature changes of the surrounding rock of the roadway at different locations during the mining of different coal seams.

Based on analysis on the experiment results, it can be seen that the temperature range of the surrounding rock was roughly symmetrically distributed when the roadway was not affected by the repeated mining in the close distance coal seams after excavation, and the temperature of the two sides of the roadway surrounding rock was higher than that of the other parts, which indicated that the stress was concentrated at the two sides of the roadway surrounding rock at this time. After the 16# coal above the roadway was mined, due to the impact of pressure relief of the upper coal seam mined, the stress transmitted from the coal pillar of the final mining line of 16# coal to the surrounding rock of the roadway decreased, resulting in a general decrease in the temperature of the surrounding rock of the roadway (to an average value of 24.1 °C). At this time, the stress concentration coefficient of the surrounding rock decreased. When the 17# coal was mined, the advance support pressure of mining of the working face directly acts on the surrounding rock, resulting in a significant increase in the stress concentration range and strength of the surrounding rock, and a sharply increased in the temperature of the surrounding rock (up to the minimum value of 25.9 °C). Compared with the 16# coal was mined, its temperature increase is 1.8 °C, and compared with non-mined, it is 1.4 °C. The temperature on the right side of the roadway (near the coal pillar) was higher than that of other parts of the surrounding rock, indicating that the stress concentration on the side near the coal pillar was high, and the stress of the surrounding rock of the roadway presents an asymmetric distribution. When the 18# coal which is below the roadway was mined, affected by the repeated mining of upper two coal seams, the temperature of the surrounding rock raised to an average value of 29.1 °C, which was 3.2 °C higher than that of the 17# coal was mined. The stress concentration of the surrounding rock of the roadway continues to rise, and the temperature of the surrounding rock of the coal pillar side was still higher than that of the other parts, indicating that the stress concentration was still in the coal pillar side. According to the analysis of Figure 12, the temperature of both sides of the roadway was generally higher than that of the roof and floor before and after the mining of the coal seams. Additionally, the temperature of the coal pillar side was always the highest compared with other parts, indicating that the stress concentration of the two sides was high, while the stress of the coal pillar side was the most concentrated.

6. Conclusions

1. It puts forward an accurate laying of model and precise excavation of roadway test method named LPDLPRM which can effectively solve the problem of accurate laying of model and precise excavation of roadway in physical similarity simulation test of roadway with special surrounding rock structures. Additionally, the neat excavation of roadway outline can be realized by this method, so that the monitoring data are accurate and reliable, which can provide reference for the physical similarity simulation test of this kind of roadway.
2. Due to the influence of repeated mining disturbance, the cracks distribution of the surrounding rock of the coal-rock rise in the close distance coal seams shows the asymmetric characteristics and it was denser in coal pillar side than the other parts. With the lower coal seam being mined, the cracks of the surrounding rock increased gradually. Under this condition, measures should be taken to strengthen the support in coal pillar side and to control the asymmetric deformation.
3. Under the influence of repeated mining in the close distance coal seams, the stress concentration of roadway surrounding rock was weakened due to the pressure relief of upper coal seam being mined. The abutment pressure increased significantly when the coal seam and the lower coal seam were mined, resulting in a sharp stress increase in the surrounding rock, and seriously influenced on stability of roadway. Therefore, in the case of repeated mining, the support of the coal-rock rise should be strengthened in the mining process of the coal seam where the roadway was located and the lower coal seam working face.

7. Patents

This paper produced a patent: A physical similarity model for roadways or tunnels with special surrounding rock structures. The patent number is ZL202122187976.2.

Author Contributions: Conceptualization, P.L. (Pengze Liu) and L.G.; methodology, P.L. (Pengze Liu) and L.G.; software, P.Z. and Y.W.; validation, G.W., L.G. and P.L. (Pengze Liu); formal analysis, P.L. (Pengze Liu); investigation, P.L. (Pengze Liu), L.G, X.K., Z.M., D.K. and S.H.; resources, L.G.; data curation, P.L. (Pengze Liu); writing—original draft preparation, P.L. (Pengze Liu); writing—review and editing, L.G.; visualization, L.G. and P.L. (Pengze Liu); supervision, P.L. (Pengze Liu); project administration, P.L. (Pengze Liu); funding acquisition, L.G., X.K. and P.L (Ping Liu). All authors have read and agreed to the published version of the manuscript.

Funding: This research was funded by the National Natural Science Foundation of China (nos. 52004073 and 52064009); the Science and Technology Support Plan of Guizhou Province (no. Qian Ke He Zhi Cheng [2021] General 400); the Science and Technology Foundation of Guizhou Province (no. Qian Ke He Ji Chu [2020] 1Y216); the Guizhou Science and Technology Plan Project (Qianke Science Foundation [2020] 1Z047); the Scientific Research Project for Talents Introduction of Guizhou University (no. Gui Da Ren Ji He Zi (2020) no. 42); the Cultivation Project of Guizhou University (no. Guidapeiyu [2019] no. 27); the Open Project Fund of Key Laboratory of Mining Disaster Prevention and Control (no. SMDPC202106) during the research.

Institutional Review Board Statement: The study did not require ethical approval.

Informed Consent Statement: Informed consent was obtained from all subjects involved in the study.

Data Availability Statement: The data presented in this study are available on request from the corresponding author.

Conflicts of Interest: The authors declare no conflict of interest.

References

1. Zhang, P.D.; Gao, L.; Liu, P.Z.; Wang, Y.Y.; Liu, P.; Kang, X.T. Study on the influence of borehole water content on bolt anchoring force in soft surrounding rock. *Shock Vib.* **2022**, *2022*, 2384626. [CrossRef]
2. Gao, L.; Zhao, S.H.; Huang, X.F.; Ma, Z.Q.; Kong, D.Z.; Kang, X.T.; Han, S. Experimental study on surrounding rock haracteristics of gateway in Panjiang mining area. *J. GZU (Nat. Sci.)* **2022**, 1–6. Available online: http://kns.cnki.net/kcms/detail/52.5002.N.20220412.0956.002.html (accessed on 13 April 2022).
3. Yu, Y.; Shen, W.L.; Guo, J. Deformation mechanism and control of lower seam roadway of contiguous seams. *J. Min. Saf. Eng.* **2016**, *33*, 49–55.
4. Chai, J.; Du, W.G.; Yuan, Q.; Zhang, D.D. Study on the mechanism of rock mass fracture around high internal pressure roadway by establishing physical similarity model. *Opt. Fiber Technol.* **2019**, *48*, 84–94. [CrossRef]
5. Guo, J.G.; Li, Y.H.; Shi, S.H.; Jiang, Z.S.; Chen, D.D.; He, F.L.; Xie, S.R. Self-forming roadway of roof cutting and surrounding rock control technology under thick and hard basic roof. Journal of China Coal Society. *J. China Coal Soc.* **2021**, *46*, 2853–2864.
6. Chen, S.J.; Wang, H.L.; Zhang, J.W.; Xing, H.L.; Wang, H.Y. Experimental study on low-strength similar-material proportioning and properties for coal mining. *Adv. Mater. Sci. Eng.* **2015**, *2015*, 696501. [CrossRef]
7. Li, L.; Wang, M.Y.; Fan, P.X.; Jiang, H.M.; Cheng, Y.H.; Wang, D.R. Strain rockbursts simulated by low-strength brittle equivalent materials. *Adv. Mater. Sci. Eng.* **2016**, *2016*, 5341904. [CrossRef]
8. Kong, H.L.; Wang, H.Z.; Gu, G.Q.; Xu, B. Application of DICM on similar material simulation experiment for rocklike materials. *Adv. Civ. Eng.* **2018**, *2018*, 5634109.
9. Mei, C.; Fang, Q.; Luo, H.W.; Yin, J.G.; Fu, X.D. A synthetic material to simulate soft rocks and its applications for model studies of socketed piles. *Adv. Civ. Eng.* **2017**, *2017*, 1565438. [CrossRef]
10. Chen, X.G.; Wang, Y.; Mei, Y.; Zhang, X. Numerical simulation on zonal disintegration in deep surrounding rock mass. *Sci. World. J.* **2014**, *2014*, 379326. [CrossRef]
11. Huang, F.; Zhu, H.H.; Xu, Q.W.; Cai, Y.C.; Zhuang, X.Y. The effect of weak interlayer on the failure pattern of rock mass around tunnel-Scaled model tests and numerical analysis. *Tunn. Undergr. Space Technol.* **2013**, *35*, 207–218. [CrossRef]
12. Sterpi, D.; Cividini, A. A physical and numerical investigation on the stability of shallow tunnels in strain softening media. *Rock Mech. Rock Eng.* **2004**, *37*, 277–298. [CrossRef]
13. Kong, D.Z.; Li, Q.; Wu, G.Y.; Song, G.F. Characteristics and control technology of face-end roof leaks subjected to repeated mining in close-distance coal seams. *Bull. Eng. Geol. Environ.* **2021**, *80*, 8363–8383. [CrossRef]
14. Lin, J.; Wang, Y.; Yang, J.H.; Wang, Z.S.; Cai, J.F. Simulation studies on stress field evolution of roadway excavation under different confining pressures. *J. China Coal Soc.* **2015**, *40*, 2313–2319.
15. Chen, X.G.; Zhang, Q.Y.; Liu, D.J.; Zhang, N.; Li, S.C. A3-D geomechanics model test study of the anchoring character for deep tunnel excavations. *Chin. Civ. Eng. J.* **2011**, *44*, 107–111.
16. Jongpradist, P.; Tunsakul, J.; Kongkitkul, W.; Fadsiri, N.; Arangelovski, G.; Youwai, S. High internal pressure induced fracture patterns in rock masses surrounding caverns: Experimental study using physical model tests. *Eng. Geol.* **2015**, *197*, 158–171. [CrossRef]
17. Yan, H.; Zhang, J.X.; Feng, R.M.; Wang, W.; Lan, Y.W.; Xu, Z.J. Surrounding rock failure analysis of retreating roadways and the control technique for extra-thick coal seams under fully-mechanized top caving and intensive mining conditions: A case study. *Tunn. Undergr. Space Technol.* **2020**, *97*, 103241. [CrossRef]
18. Hui, G.L.; Niu, S.J.; Jing, H.W.; Wang, M. Physical simulation on deformation rules of god-side roadway subjected to dynamic pressure. *J. Min. Saf. Eng.* **2010**, *27*, 77–81+86.
19. Mishra, S.; Kumar, A.; Rao, K.S.; Gupta, N.K. Experimental and numerical investigation of the dynamic respone of tunnel in soft rocks. *Structures* **2021**, *29*, 2162–2173. [CrossRef]

20. Shan, R.L.; Huang, B.; Zheng, Y.; Kong, X.S.; Zhang, S.P.; Zhang, L.Z. Development of similar simulation equipment for roadway support subjected to vertical dynamic loads. *Chin. J. Geotech. Eng.* **2018**, *40*, 1163–1173.
21. Chang, J.C.; Li, D.; Xie, T.F.; Shi, W.B.; He, K. Deformation and failure characteristics and control technology of roadway surrounding rock in deep coal mines. *Geofluids* **2020**, *2020*, 8834347. [CrossRef]
22. Oh, D.W.; Lee, Y.J. Analysis of pile load distribution and ground behaviour depending on vertical offset between pile tip and tunnel crown in sand through laboratory model test. *J. Korean Tunn. Undergr. Space Assoc.* **2017**, *19*, 355–373. [CrossRef]
23. Li, Y.H.; Liu, D.Z.; Yang, S. Development and application of physical simulation test system for small and medium-sized tunnels based on biaxial motor loading. *Chin. J. Geot. Eng.* **2020**, *42*, 1556–1563.
24. Yuan, Z.G.; Shao, Y.H.; Zhu, Z.H. Similar Material Simulation Study on Protection Effect of Steeply Inclined Upper Protective Layer Mining with Varying Interlayer Distances. *Adv. Civ. Eng.* **2019**, *2019*, 9849635. [CrossRef]
25. Wang, Z.Y.; Dou, L.M.; He, J. Failure characteristics of rock bursts in steeply-inclined extra-thick coal seams under static-dynamic coupling loading. *J. Min. Saf. Eng.* **2021**, *38*, 886–894.
26. Idinger, G.; Aklik, P.; Wu, W.; Borja, R.I. Centrifuge model test on the face stability of shallow tunnel. *Acta Geotech.* **2011**, *6*, 105–117. [CrossRef]
27. He, M.C.; Gong, W.L.; Zhai, H.M.; Zhang, H.P. Physical modeling of deep ground excavation in geologically horizontal strata based on infrared thermography. *Tunn. Undergr. Space Technol.* **2010**, *25*, 366–376. [CrossRef]
28. Li, H.L.; Bai, H.B.; Ma, D.; Tian, C.D.; Zhang, Q. Physical simulation testing research on mining dynamic loading effect and induced coal seam floor failure. *J. Min. Saf. Eng.* **2018**, *35*, 366–372.
29. Ren, Y.F.; Ning, Y.; Qi, Q.X. Physical analogous simulation on the characteristics of overburden breakage at shallow longwall coalface. *J. China Coal Soc.* **2013**, *38*, 61–66.
30. Shi, X.S.; Jing, H.W.; Zhao, Z.L.; Gao, Y.; Zhang, Y.C.; Ruodi, B. Physical experiment and numerical modeling on the failure mechanism of gob-side entry driven in thick coal seam. *Energies* **2020**, *13*, 5425. [CrossRef]
31. Berthoz, N.; Branque, D.; Subrin, D.; Wong, H.; Humbert, E. Face failure in homogeneous and stratified soft ground: Theoretical and experimental approaches on 1g EPBS reduced scale model. *Tunn. Undergr. Space Technol.* **2012**, *30*, 25–37. [CrossRef]
32. Feng, C.; Dong, S.; Lai, X.P.; Chen, J.Q.; Cao, J.T.; Shan, P.F. Study on rule of overburden failure and rock burst hazard under repeated mining in fully mechanized top-coal caving face with hard roof. *Energies* **2019**, *12*, 4780.
33. Xue, J.H.; Wang, H.P.; Zhou, W.; Ren, B.; Duan, C.R.; Deng, D.S. Experimental research on overlying strata movement and fracture evolution in pillarless stress-relief mining. *Int. J. Coal Sci. Technol.* **2015**, *2*, 38–45. [CrossRef]
34. Yang, K.; He, X.; Dou, L.T.; Liu, W.J.; Sun, L.; Ye, H.S. Experimental investigation into stress-relief characteristics with upward large height and upward mining under hard thick roof. *Int. J. Coal Sci. Technol.* **2015**, *2*, 91–96. [CrossRef]
35. Zhang, S.K.; Lu, L.; Wang, Z.M.; Wang, S.D. A physical model study of surrounding rock failure near a fault under the influence of footwall coal mining. *Int. J. Coal Sci. Technol.* **2021**, *8*, 626–640. [CrossRef]
36. Liu, P.Z.; Gao, L.; Zhang, P.D.; Wu, G.Y.; Wang, C.; Ma, Z.Q.; Kong, D.Z.; Kang, X.T.; Han, S. A case study on surrounding rock deformation control technology of gob-side coal-rock roadway in inclined coal seam of a mine in Guizhou, China. *Processes* **2022**, *10*, 863. [CrossRef]
37. Gao, L.; Liu, P.Z.; Zhang, P.D.; Wu, G.Y.; Kang, X.T. Influence of fracture types of main roof on the stability of surrounding rock of the gob-side coal-rock roadway in inclined coal seams and its engineering application. *Goal Geol. Explor.* **2022**, 1–8. Available online: https://kns.cnki.net/kcms/detail/61.1155.p.20220416.0838.002.html (accessed on 5 April 2022).
38. Kong, D.Z.; Xiong, Y.; Cheng, Z.B.; Wang, N.; Wu, G.Y.; Liu, Y. Stability analysis of coal face based on coal face-support-roof-system in steeply inclined coal seam. *Geomech. Eng.* **2021**, *25*, 233–243.
39. Ji, Y.M. Study on Infrared Characteristics of Bolt and Rock in Condition of Loading. *Spectrosc. Spectr. Anal.* **2010**, *30*, 659–662.
40. Toubal, L.; Karama, M.; Lorrain, B. Damage evolution and infrared thermography in woven composite laminates under fatigue loading. *Int. J. Fat.* **2006**, *28*, 1867–1872. [CrossRef]
41. Yang, Z.; Qi, Q.J.; Ye, D.D.; Li, X.; Luo, H. Variation of internal infrared radiation temperature of composite coal-rock fractured under load. *J. China Coal Soc.* **2016**, *41*, 618–624.
42. Peng, Y.Y.; Lin, Q.C.; He, M.C.; Zhu, C.; Zhang, H.J.; Guo, P.F. Experimental study on infrared temperature characteristics and failure modes of marble with prefabricated holes under uniaxial compression. *Energies* **2021**, *14*, 713. [CrossRef]
43. Ma, L.Q.; Li, Q.Q.; Cao, X.Q.; Zhou, T. Variation characteristics of internal infrared radiation temperature of coal-rock mass in compression process. *J. China Univ. Min. Technol.* **2013**, *42*, 331–336.
44. Meola, C. A new approach for estimation of defects detection with infrared thermography. *Mater. Lett.* **2007**, *61*, 747–750. [CrossRef]
45. Liu, S.J.; Wu, L.X.; Wu, H.P.; WU, Y.H.; Cheng, T.; Li, G.H. Quantitative studies of infrared radiation dark mineral uniaxial loading process. *Chin. J. Rock Mech.* **2002**, *21*, 1585–1589.

MDPI

St. Alban-Anlage 66

4052 Basel

Switzerland

Tel. +41 61 683 77 34

Fax +41 61 302 89 18

www.mdpi.com

Energies Editorial Office

E-mail: energies@mdpi.com

www.mdpi.com/journal/energies